License Plate Readers for Law Enforcement

Opportunities and Obstacles

Keith Gierlack, Shara Williams, Tom LaTourrette,
James M. Anderson, Lauren A. Mayer, Johanna Zmud

The research described in this report was sponsored by the National Institute of Justice and was conducted in the Safety and Justice Program within RAND Justice, Infrastructure, and Environment.

This project was supported by Award No. 2010-IJ-CX-K007, awarded by the National Institute of Justice, Office of Justice Programs, U.S. Department of Justice. The opinions, findings, and conclusions or recommendations expressed in this publication are those of the authors and do not necessarily reflect those of the Department of Justice.

Library of Congress Cataloging-in-Publication Data is available for this publication.

ISBN: 978-0-8330-8710-2

The RAND Corporation is a nonprofit institution that helps improve policy and decisionmaking through research and analysis. RAND's publications do not necessarily reflect the opinions of its research clients and sponsors.

Cover image: Police car equipped with mobile ANPR produced by ELSAG North America (Mobile Plate Hunter 900). Kafuffle via Wikimedia; CC BY 2.0.

RAND OFFICES

SANTA MONICA, CA • WASHINGTON, DC

PITTSBURGH, PA • NEW ORLEANS, LA • JACKSON, MS • BOSTON, MA

CAMBRIDGE, UK • BRUSSELS, BE

www.rand.org

Preface

Since the use of license plate reader (LPR) technology is relatively new in the United States, opportunities and obstacles in its use in law enforcement are still under exploration. As the technology spreads, however, law-enforcement agencies, particularly those considering investing in an LPR system and other organizations focused on the information technology needs of law enforcement, may find the material in this report helpful. It provides an in-depth examination of the range of ways in which license plate scanners are used; the benefits and limits of LPR systems; and emerging practices for system operation. The RAND Corporation's research approach, exploratory interviews with law-enforcement personnel, sought to gather information not just from police officers but also from the diverse people responsible for installing, maintaining, and operating the systems. This method allowed RAND to thoroughly characterize and examine license plate scanner issues to add to the knowledge base. The interviews explored salient issues concerning system implementation, funding, case uses, field procedures, technology issues, data retention policies, and privacy concerns. RAND believes these findings overall will add value to the discussion on this technology's utility. This work was sponsored by the National Law Enforcement and Corrections Technology Center of Excellence on Information and Geospatial Technology at the National Institute of Justice (NIJ). It should be of interest to law-enforcement personnel at all levels and is one in a series of NIJ-sponsored resources for police departments.

The RAND Safety and Justice Program

The research reported here was conducted in the RAND Safety and Justice Program, which addresses all aspects of public safety and the criminal justice system, including violence, policing, corrections, courts and criminal law, substance abuse, occupational safety, and public integrity. Program research is supported by government agencies, foundations, and the private sector.

This program is part of RAND Justice, Infrastructure, and Environment, a division of the RAND Corporation dedicated to improving policy and decisionmaking in a wide range of policy domains, including civil and criminal justice, infrastructure protection and homeland security, transportation and energy policy, and environmental and natural resource policy.

Questions or comments about this report should be sent to the project leader, Johanna Zmud (jzmud@rand.org). For more information about the Safety and Justice Program, see http://www.rand.org/safety-justice or contact the director at sj@rand.org.

Contents

Figures and Tables

Figures

Tables

Summary

Financed largely by federal and state grants, law-enforcement agencies across the country have quickly been adopting a new technology to combat auto theft and other crimes: automated license plate readers. The license plate reader (LPR) system, which has portable and fixed units, can capture the image of a passing vehicle and compare its plates against official "hotlists" that show an array of infractions in which it may be involved or reasons why it may be of interest to authorities. After an alert is issued, the officer can then investigate the license plate of interest and decide whether to take further action.

Not only can the system alert officers with this actionable information, but it can also provide photos of the car or truck in question. This technology captures images of exponentially more vehicles than any single officer possibly can, and the information generated may also be stored in databases, allowing investigators potentially to analyze these data for more complex and open investigations of serious crimes.

Although this technology has spread rapidly in agencies nationwide, the dearth of deeper research means that police and policymakers may not possess sufficient, credible information to best assess the opportunities and challenges of these systems. Their initial grant funding will expire soon, meaning that already budget-strapped departments will be forced to make tough decisions about buying more such equipment, replacing or upgrading it, maintaining it, and expanding its use.

Further, as investigators see the potential of querying past records that the technology produces—for example, to track crime patterns or trends or in pursuit of criminal suspects—privacy advocates have grown wary, and some lawmakers have imposed restrictions or outright bans on these systems or aspects of their use.

To forestall and address public concerns, agencies using the systems have struggled with new, complicated matters of technology management, including how and with whom they should share data, who may access it, and how long they might keep the information. Seasoned police officers, with major help from system administrators with information technology (IT) backgrounds, have had to figure out how to install, maintain, and repair cameras, computers, and servers and learn whether their high-tech systems will be compatible with installations of other departments in the area. Agencies have scrutinized their personnel and staffing and have developed system trainings for patrol officers and investigators. Those that have grappled with the technology's issues and know the systems best are already inching their way to resolutions.

This RAND Corporation study is grounded in a survey of existing literature, case study interviews of users steeped in the systems, and an analysis and provision of common themes and lessons learned. It seeks to assist those who are zeroing in on this technology by offering more comprehensive research and information than heretofore available. The way that LPR technology will advance, say its proponents, is for its users to cooperate and share experiences, ideas, and knowledge about best practices.

The study consolidates existing and new information about LPR to offer law-enforcement leaders and policymakers several paths forward for what many see as a crime-fighting tool with great potential. The research team concludes that this technology, while promoted initially as a tool to assist in the area of auto theft—both to spot and recover stolen vehicles—can be used in many more ways that law-enforcement agencies have only begun to discover and try. The systems can be used in both reactive and analytic fashion; they possess significant potential to act as "force multipliers" because they can scan and process so many more vehicles and license plates than individual officers ever can. While they can produce an information overload if not adjusted and regulated, especially by officers using them in the field, the systems can be useful and reactive, scanning plates, matching them against "hotlists" of an array of infractions, and generating valuable alerts and collateral information about vehicles of interest. This can save investigative time and contribute to the safety of law-enforcement personnel making stops on the street. Systems also can build revenue when officers use them to help track down offenders, resulting payments of fines owed.

This technology, however, has not reached its full crime-fighting potential, and RAND has found that multiple factors constrain it. Law-enforcement officers have not taken maximum advantage yet of the analytic capacity of the volumes of information created by plate-reader systems and stored in computer databases. These data already allow them to process only partial plate information to pare lists quickly of many possible vehicles of interest. The systems' data could help officers discover crime trends and hot spots of criminal activity. It could help them test the alibis of burglars, killers, and other criminals. It could support police missions to combat drug cartels and terrorist groups.

But agencies lack the time, training, knowledge, and experience to work with this voluminous information and struggle even now just to share and to store it. Technology to deal with these issues can be daunting; because many departments tapped grants that once were plentiful to buy their systems, they did not plan for repairing, maintaining, upgrading, and expanding their plate reader systems, given their high-tech, weather-exposed, heavily used equipment—nor do they have money to do so. Departments also wrestle with ways to cooperate with each other, not just to share data but on fundamental matters of their systems' interoperability. Meantime, the LPR technology, RAND found, works best if it taps robust streams of information fed in from many different agencies, contributing different lists of license plates in varied infractions or of interest for many reasons.

Even as agencies increase their plate data and data storage, their retention and use of this information raises increasing privacy concerns about state surveillance. Many agencies lack retention, access, and other important policies or guidelines on this sensitive information. And RAND researchers see with some certainty more litigation in this area, up to the U.S. Supreme Court, as well as potential legislative action at the local, state, and national levels.

Still, champions of this technology exist at many levels, from tech-savvy officers who use it every day, to system administrators who make it run, to chiefs who promote it, to other officials and policymakers. These proponents will need to promote national standards and models covering the systems' equipment, data records, information sharing, access, and retention. They will need not only to make the case for new, sound uses of the technology but also for an array of supports for it—including financing, training, staffing, and potential new tools for considering costs, benefits, outcomes, and risks.

Our research points to three major issues and several immediate policy steps for agencies using LPR to consider. First, as noted, agencies are often constrained from realizing the full reactive and analytic capabilities of LPR systems. In the case of reactive uses, the lack of access to all available data sources, from stolen vehicle and warrant hotlists to Department of Motor Vehicles data, decreases the utility of the system by narrowing the range of uses. In the case of analytic uses, the lack of available resources and personnel to support the analytic potential of the technology is of particular concern, especially for smaller agencies. Departments may also benefit from being able to retain data for a long time, allowing investigators to identify trends or patterns.

Second, privacy concerns can arise in use of stored data. Law-enforcement storing of plate data for use in later investigations raises serious concerns among privacy advocates. Advances in Fourth Amendment doctrine have fallen behind rapid progress in modern information technology. Recent U.S. Supreme Court decisions on individuals' locational privacy have conflicted with or left key questions unanswered. Many issues regarding use of this technology remain to be resolved by the courts. In the meantime, different agencies have improvised as best they can.

Third, the utility of LPR systems depends on how well they are set up to receive and share data. Our research identified three challenges to interagency data sharing: data availability, developing and signing administrative sharing agreements, and technical interoperability of systems, including standardization of records. Given that two large vendors dominate the market, agencies in neighboring jurisdictions may operate compatible systems—if they coordinate their purchases beforehand. Even if systems are compatible, data cannot be shared without standardization.

These and other issues point to several recommendations for enhancing use of LPR technology. They include the following:

- Estimating and securing necessary funding for the entire lifecycle of this technology. Funding should take into account not only acquisition costs but also costs of staffing and maintenance.
- Ensuring that sufficient IT infrastructure is in place to handle different types of data promptly and frequently. The structure must be able to handle data coming in and going out of the system so that departments can cooperate effectively.
- Developing policies for system data use, access, and storage. Clear guidelines on use, storage, and retention of data will help personnel consistently and legally use the technology in ways that are most valuable. Agencies lacking policies may wish to consult department counsel and local and state officials for guidance. Agencies should also be prepared to revise policies as public or political support or concern over use of this technology change over time.
- Integrating LPR systems into daily agency operations and learning from other agencies how to expand their use to more analytical operations. Agencies should consider this technology as more than a "stolen vehicle finder." To improve its cost-benefit ratio, agencies should consider all opportunities for use of this technology.
- Developing model memoranda of understanding for agencies to share LPR data. Existing agreements of agencies already sharing such data may be adapted to other circumstances.
- Identifying tradeoffs between privacy rights and law-enforcement utility. Agencies may use risk-taxonomy technical standards based on information in this report to identify and estimate exposure to privacy risks and specify potential tradeoffs with law-enforcement utility.

Acknowledgments

This work would not have been possible without the assistance of many people inside and outside of RAND. We are especially grateful for assistance provided by the professional staff of seven police departments in the United States. We do not name them here due to confidentiality of our sources. We thank our anonymous National Institute of Justice (NIJ) reviewers and our RAND colleague Joel Predd for their very helpful comments on the draft, which have resulted in a better report. Debra Knopman, also a RAND colleague, made valuable comments as well. We also thank Clifford Grammich and Craig Masuda at RAND for their value-added editing of our manuscript. Finally, we thank Steve Schuetz at NIJ for his helpful insights and strong support of this project.

Abbreviations

ACLU	American Civil Liberties Union
DMV	Department of Motor Vehicles
DUI	driving under the influence
FBI	Federal Bureau of Investigation
FTE	full-time equivalent
GPS	Global Positioning System
IACP	International Association of Chiefs of Police
IT	information technology
LEMAS	Law Enforcement Management and Administrative Statistics
LPR	license plate reader
MoU	memorandum of understanding
NCIC	National Criminal Information Center
NICB	National Insurance Crime Bureau
NIJ	National Institute of Justice
PII	personally identifiable information

CHAPTER ONE

Introduction

Background

In 2011, New York City detectives arrested Luis Zeldon in the killing of Andy Herrera in Queens. A primary analytic tool used in this homicide investigation and arrest was a system of license plate readers (LPRs) located throughout the city. After Zeldon was identified as the main suspect, police turned to the city's LPR database to help identify his possible hideouts, based on location information on his vehicle. A digital map created from the system's records helped police identify an address near the victim that was associated with Zeldon. This led officers directly to his location, where he was found hiding in a closet.[1] He later pleaded guilty to manslaughter in Herrera's slaying.[2]

Such incidents have become more frequent as plate reading technology has spread across the United States. It was launched in the United Kingdom in the 1990s as part of a comprehensive network of counterterrorism surveillance assets. The technology migrated to the United States, where its extensive use began in the last decade. It has spread to hundreds of U.S. law-enforcement agencies at the local, state, and federal levels.

LPR systems typically pair infrared and visible-light cameras to scan surrounding area for license plates. The infrared camera, with optical character recognition software, can identify license plates and "read" plate characters; the vehicle and plate are both photographed. License plate information is then compared to a database or "hotlist" of plates connected with criminal activity to determine if the scanned license plate is of interest to lawenforcement. If a match is detected, the system alerts the officer and, in some cases, always displays a photo so an operator quickly can identify the suspect vehicle.

---

1 Al Baker, "Camera Scans of Car Plates Are Reshaping Police Inquiries," *New York Times*, 11 April 2012.

2 Office of Richard A. Brown, District Attorney, Queens County, New York, "Queens Man Pleads Guilty to Manslaughter in Death of Roommate's Friend," 4 November 2011.

The LPR systems, while primarily used to detect stolen vehicles and plates, are increasingly being tapped for a variety of investigations.[3] Authorities can mine databases to determine vehicles in the vicinity of a crime scene, to provide photos of those vehicles to confirm suspect alibis, and even to analyze crime patterns.[4] In response to the LPR systems' perceived added value, many agencies have taken advantage of federal and state grants to buy and expand this technology, with some larger departments deploying numerous cameras and extensive databases.

These LPR systems may consist of fixed, portable, and mobile cameras coupled with searchable databases. Fixed cameras scour a specific geographic area and transmit data back to administrators to analyze. These cameras can be installed in high-crime or high-traffic areas or on main thoroughfares to act as both added eyes for police and a potential deterrent to criminal activity. Portable systems often are housed in inconspicuous trailers that stay put for a time and then can be relocated. Mobile cameras, affixed to police cruisers or handheld, can help authorities patrol operational areas or target operations in high-crime areas. Databases with downloaded plate information can be searched using even partial license numbers; investigators can retrieve time and geospatial information. These LPR systems provide police with more investigative tools for a variety of crimes.

Although law enforcement widely views the technology as a positive development (see, e.g., Roberts and Casanova, 2012), many privacy advocates have challenged the practice of storing plate data not associated with a specific crime.[5] System databases record every license plate read, even those not on a hotlist, in a searchable format for later use. As a result, many privacy advocates say law-enforcement agencies could track individuals by the movement of their vehicles. The LPR systems' data retention, or the length of time records are kept in a database, has become contentious. Some police departments lack clear guidance on storing plate data, leaving privacy advocates to fear it can be kept and retrieved indefinitely. In response, some privacy advocates, departments, and lawmakers have moved to codify police procedures on recording these data; some have banned the technology's use outright.

---

[3] Cynthia Lum, Linda Merola, Julie Willis, and Breanne Cave, *License Plate Recognition Technology (LPR): Impact Evaluation and Community Assessment*, Center for Evidence-Based Crime Policy, George Mason University, Fairfax, VA, 2010; David J. Roberts and Meghann Casanova, *Automated License Plate Recognition (ALPR) Use by Law Enforcement: Policy and Operational Guide,* International Association of Chiefs of Police, Washington, DC, 2010.

[4] Eric Roper, "Cops Move to Protect License Plate Data," *Minneapolis Star Tribune*, 4 November 2012.

[5] American Civil Liberties Union (ACLU), *You Are Being Tracked: How License Plate Readers Are Being Used to Record Americans' Movements*, New York, July 2013.

Rationale for NIJ-Sponsored Research

Since the use of license plate reader technology is relatively new in the United States, its opportunities, strengths, and weaknesses in law enforcement are still under exploration. The limited research on its use focuses primarily on aggregated survey data on ownership and use trends among law-enforcement agencies, the benefits of its singular application in auto theft cases, or advocacy group perspectives related to privacy concerns. As the technology spreads, law-enforcement agencies, particularly those considering investing in a system and other organizations focused on law enforcement's information technology (IT) needs, may find it helpful to have an in-depth examination of the range of ways in which license plate scanners are used, the benefits and limits of LPR systems, and emerging practices for system operation. This RAND report, sponsored by the National Institute of Justice (NIJ), seeks to provide such an in-depth examination. It uses structured interviews exploring a wide range of topics on plate scanner technology, including the types of investigations that benefit most from it, interoperability among systems, funding sources, and actions taken to alleviate privacy concerns.

RAND's research approach—exploratory interviews with law-enforcement personnel—sought to gather information not only from police officers but also from the diverse people responsible for installing, maintaining, and operating the LPR systems. Administrators, analysts, grant writers, and senior officials with intimate knowledge of the systems' benefits and challenges had their say. This method allowed RAND to thoroughly characterize and examine license plate scanner issues to add to the knowledge base. RAND employed multiple criteria derived from Law Enforcement Management and Administrative Statistics (LEMAS) data, U.S. Census data, Uniform Crime Reporting (UCR) statistics, and independent research to select seven case study agencies. Case study interviews explored salient issues on LPR systems' implementation, funding, types of investigations, field procedures, technology issues, data retention policies, and privacy concerns. Our findings confirmed some generally held beliefs, while challenging others. RAND believes that these findings overall will add value to the discussion on this technology's utility. To encourage interviewees' frank discussion of issues, RAND does not identify departments participating in this report.

The comprehensive nature of this research was designed to bring to the surface aspects of LCR technology not reported in media or academic circles. The authors hope the information RAND provides here will assist in investment decisions among law-enforcement agencies and in formulating relevant questions for follow-up research by NIJ or others.

Key Influencing Factors in LPR Utility

RAND undertook the study of these LPR systems to enhance understanding of the benefits, challenges, and practices of this fast diffusing technology. From the beginning, we focused on developing a typology of major-use cases to explore how this technology's utility might vary. This is relevant because the technology requires substantial investment in money and resources for any department. Law-enforcement decision-makers, therefore, want to know the types of investigations for which it provides the most utility.

We identified two use types: (1) reactive, where this technology provided instantaneous information to officers, and (2) analytic, where officers or analysts could mine information from LPR system databases. But as our research progressed, we found that key issues and themes associated with benefits, challenges, and practices were not use-case dependent.

Two factors appeared to influence the utility of these LPR systems: database access and data retention policies. A key to the technology's effectiveness was access to databases of related, criminal information (such as stolen vehicle and warrant hot-lists) and vehicle information (such as state Department of Motor Vehicles [DMV] data). The length of time departments retained data also was critical to provision of analytical benefits. Departments that kept data for considerable periods—months or years—could mine it to provide information to investigators about ongoing cases or to identify crime trends and patterns. Departments that erased data quickly, in days or weeks, lacked this capability. Therefore, while categorizations such as reactive and analytic were still relevant to framing our research, the more important finding was that LPR systems can be beneficial in providing assistance in any type of investigation, provided the necessary data are available to support the LPR systems. Systems with the most database access and longest retention policies are the most beneficial because they can provide the greatest number of alerts across law-enforcement activities; records of both alert and non-alert reads are kept long enough to provide utility for more long-term investigations or crime analysis. Inversely, systems that have limited access to data and discard records in a short time period limit the types of uses for which LPR is relevant and do not provide the ability to query past records for ongoing or past investigations. Figure 1.1 illustrates how data access and retention policies influence the systems' utility.

Report Structure

In Chapter Two, we review the literature, looking at the bounds in which license plate readers operate and setting context for RAND's findings. Chapter Three describes the methodology used to select law-enforcement agencies for case study. Chapter Four

Figure 1.1
Identifying the Usefulness of LPR Systems

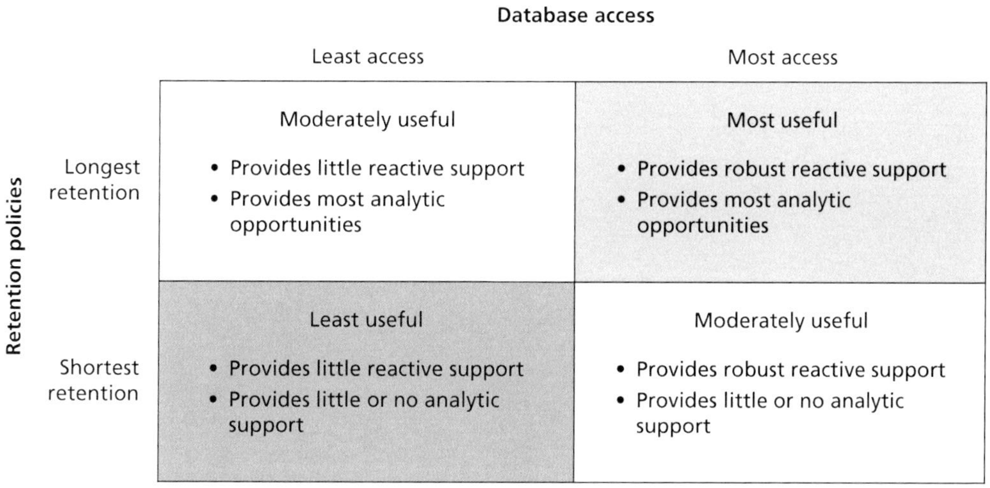

Database access

Retention policies	Least access	Most access
Longest retention	Moderately useful • Provides little reactive support • Provides most analytic opportunities	Most useful • Provides robust reactive support • Provides most analytic opportunities
Shortest retention	Least useful • Provides little reactive support • Provides little or no analytic support	Moderately useful • Provides robust reactive support • Provides little or no analytic support

RAND *RR467-1.1*

analyzes license plate readers in various operational environments, comparing departments on system structure, policies, data retention, interoperability, and information sharing. Chapter Five explores key issues surrounding privacy concerns. Chapter Six presents our case study findings, putting them forward as themes that emerged. Chapter Seven offers lessons learned and recommendations for the technology's use from the case study agencies. We offer concluding observations and recommendations in Chapter Eight. Appendix A describes how each case study department put in place, operates, and maintains its LPR systems and details our interview questions. Appendix B provides the protocol used for the case study interviews.

CHAPTER TWO

Framing the LPR Environment

RAND conducted a literature review to understand current LPR use, to frame its operating environment, and to guide our interviews with law-enforcement participants. This chapter also covers descriptions of the LPR systems' case uses, costs, benefits, and challenges, and privacy issues. LPR technology was invented in the mid-1970s in the United Kingdom. In the early 1990s, Britain responded to terrorism by the Provisional Irish Republican Army by establishing a surveillance and security network around London called the "ring of steel." It mainly relied on closed circuit TV cameras to increase police awareness in the city. Starting in 1997, officials installed license plate reading cameras at the entrances to this network to give police advance notice when vehicles of interest entered.[1] Then, in 2002, police in nine different jurisdictions in the UK began testing dedicated intercept teams, using these cameras to target criminals as part of "Project Laser,"[2] which was aimed at criminals on the roads, allowing police to intercept them and terrorist elements. The project was deemed a success, with more than 46,000 arrests resulting from plate recognition.[3] As a result of Project Laser, the Home Office invested significantly to develop a National ANPR[4] Data Centre and a Back Office Facility to fully exploit the sizable amount of data collected by license cameras across the UK.[5]

The technology migrated to the United States, with the U.S. Border Patrol first employing system cameras at border crossings in 1998.[6] The technology has since become a tool of interest to law-enforcement agencies, with several recent surveys indicating a substantial number of agencies employing it. According to the LEMAS 2007 report, 19 percent of almost 900 law-enforcement agencies responding to the LEMAS

---

[1] Roberts and Casanova, p. 3.

[2] Roberts and Casanova, p. 4.

[3] Roberts and Casanova, p. 4.

[4] ANPR stands for Automated Number Plate Reader.

[5] Roberts and Casanova, p. 5.

[6] Gina Harkins, "License Plate Recognition Databases—Good for Cops or Invasion of Privacy?" *Medill National Security Zone*, 11 May 2011.

survey employed LPR.[7] The vast majority of these agencies, 70 percent or so, were larger and employed more than 500 officers. In 2010, a George Mason University survey found 20 percent of 169 responding agencies used LPR systems.[8] A report commissioned by the International Association of Chiefs of Police, published in 2012, found that 23 percent of 305 departments surveyed had them.[9] A 2012 survey by the Police Executive Research Forum noted that 71 percent of responding departments used the technology; 85 percent planned to buy the equipment or expand their systems in the next five years.[10] This indicates that a significant proportion of departments use plate readers.

Major Uses

Although LPR systems were first used by the British to give police extra eyes to counter terrorism in London, authorities quickly realized that the technology could be applied in many law-enforcement activities to combat auto theft and recover stolen plates and associated vehicles; find DMV violations, such as expired plates and insurance violations; track down parking scofflaws; assist in investigations of violent crimes, including homicides and gang activity; and track vehicles of known criminals. Many law-enforcement officers consider the technology an added, effective tool to combat crime.

Its uses fall into two general categories: reactive and analytic. Reactive uses involve instant alerts, in which a plate is read that is associated with a criminal hotlist. The system alerts the officer, who can pursue action against that plate in real time. The most common reactive uses are the identification of stolen vehicles, DMV violations such as expired registrations, or criminal activity associated with warrant hotlists. LPR can also be used in a reactive manner for more serious or longer-term criminal activity, such as murder, narcotics, or kidnapping, if a license plate is identified and added to a relevant hotlist. This category generally presents the fewest concerns regarding privacy issues because each relevant license plate is directly associated with the possibility of criminal activity.

Analytic uses involve the exploitation of the database of stored LPR reads to benefit ongoing investigations. When a crime occurs, an investigator can mine the database for license plates in the area of the crime for possible leads. Murder, bank robbery, Amber Alerts, and drug investigations are all examples of LPR case uses that can ben-

---

[7] *Law Enforcement Management and Administrative Statistics*, Bureau of Justice Statistics, Office of Justice Programs, United States Department of Justice, Washington, DC, 2007.

[8] Lum et al., p. 19.

[9] Roberts and Casanova, p. ix.

[10] Police Executive Research Forum, *How Are Innovations in Technology Transforming Police?* Critical Issues in Policing Series, 2012, p. 31.

efit from analyzing a database of LPR reads. Additionally, the database can be used to search for license plates known to be associated with criminal activity, such as gang or drug networks, and used to help determine areas of criminal activity. This category presents the most privacy concerns because it exploits a database of stored reads, most of which are not associated with criminal activity.

Reactive Policing

LPR technology first was introduced in the United States, ostensibly to combat auto theft.[11] The technology seemed ideal for this purpose because it could scan licenses exponentially faster than police officers could and then compare plates to police databases. Units went into areas known for high car thefts and quickly provided police a way to efficiently target their prevention and apprehensions. Recent studies show that between 83 percent[12] and 91 percent[13] of law-enforcement agencies use their LPR capabilities for auto theft investigations. A study conducted by the Ohio State Police in 2005 found that adding this equipment to turnpike entryways increased stolen vehicle recovery over the same time period previous year by approximately 50 percent.[14]

The technology can also work with DMV data in reactive applications to target vehicle and traffic violations. These LPR systems, populated by records obtained from DMV databases, can identify stolen and expired license plates and auto insurance and inspection violations. In a George Mason University survey, 40 of 134 responding departments used license plate readers to identify motor vehicle violations.[15] It also can assist municipalities in collecting parking fines. In the first month of operations in Boulder, Colorado, police with the system caught more than one hundred parking scofflaws who owed more than $20,000 to the city.[16]

The technology can also identify persons of interest if the system is linked to relevant databases. It can be especially useful in flagging and arresting wanted suspects. A police officer in Fairfax County, Virginia, pulled over a vehicle on an LPR alert for a suspended plate. After running the name of the vehicle's owner through another database, the officer discovered the driver was wanted on an outstanding felony drug warrant.[17]

---

[11] Michael Ferraresi, "License-Plate Scanning Catching Crooks, Raising Privacy Worries," *AZCentral.com*, 23 November 2008.

[12] Roberts and Casanova, p. 17.

[13] Lum et al., p. 21.

[14] *Automated Plate Reader Technology,* Ohio State Highway Patrol Planning Services Section Research and Development, Columbus, OH, 2005, Abstract.

[15] Lum et al., p. 21.

[16] Heath Urie, "Boulder License-Plate Scanner Busts 101 Parking Scofflaws in 1st Month," *Boulder County News*, 30 June 2011.

[17] Justin Jouvenal, "License Plate Reader Helps Nab Alleged Drug Dealer," *Washington Post*, 13 July 2011.

Analytic Uses

LPR's analytic uses involve searching its databases for license plates of interest in a criminal investigation. Because records of license plate scans are sometimes stored in databases, for a year or two, depending on departmental policies, officers can retrieve license plate–related information from crime scenes, increasing their chances to solve a crime. Since system data provide both geographic and time information, this tool can be employed in a variety of investigations, including drugs, homicides, burglaries, and gang activities.

These records can determine what vehicles are in the vicinity of a crime scene,[18] assisting investigators in creating lists of possible suspects or witnesses.[19] If, say, an investigator has license plate numbers of known drug traffickers, a database search could reveal areas of interest that could offer leads on drug houses or other sites of trafficking activity. License plate readers can yield clues of vehicles in the area where bank robberies are committed.[20] If a witness provides police only with partial license numbers of a vehicle involved in a crime, a "wild card" search of those numbers and letters can help police narrow suspect vehicles.[21]

Much of the reporting on the technology's analytical uses focuses on its assistance in specific, continuing investigations. But these data also can be a boon in intelligence-led policing, with authorities analyzing records in historical databases to identify crime patterns and trends. This could help them find "chop shops" for stolen vehicles. As M. Murat Ozer wrote in his dissertation about the Cincinnati Police Department: "Specifically, LPR systems vastly increase the data collection capabilities of police departments, providing them with a greater knowledge base from which to develop new, data-driven strategies."[22] Ozer continued:

> Another aspect of LPR technology is the collection of timely and accurate data about criminals and non-criminals. This feature of LPR technology provides invaluable information for intelligence-led policing; with timely and accurate data, police departments can trace the movement of criminal vehicles from one place to another. Based on this, crime patterns of a city can be mapped, which in turn will enable police departments to implement prevention efforts.[23]

---

[18] Patrick Barnard, "Fairfield Police Defend Use of License Plate Readers," *Fairfield Patch*, 3 June 2012.

[19] Roper, "Cops Move to Protect License Plate Data."

[20] Ferraresi.

[21] Eric Roper, "St. Paul Meets Minneapolis on Vehicle Tracking Data Retention," *Minneapolis Star Tribune*, 14 November 2012.

[22] M. Murat Ozer, *Assessing the Effectiveness of the Cincinnati Police Department's Automatic License Plate Reader System within the Framework of Intelligence-Led Policing and Crime Prevention Theory*, dissertation submitted to the School of Criminal Justice, College of Education, Criminal Justice, and Human Services, University of Cincinnati, 7 July 2011, p. 3. Automated License Plate Reader (ALPR) is another acronym for LPR technology.

[23] Ozer, p. 109.

Much interest in these systems focuses on direct utility of their data in the field. There also are indirect applications, such as the potential to uncover trends in criminal activity so that police can target resources on high-gain operations.

It is important to note that during the course of our research, RAND did not find evidence that this technology is ineffective in any specific type of use. It appears to provide utility to any type of investigation if supported by appropriate data access and retention policies. The greater the data access and the longer the retention policy, the more useful LPR can be to police, as noted earlier in this report; less access to fewer databases and short retention inhibit the systems' capacity to help police.

Cost Considerations

LPR costs can be divided into those for initial implementation and those at the back end. Initial system costs include purchasing cameras, installation, system integration, and training. Back-end costs include maintaining the network and storing data. For departments on a tight budget already, funding a system can be a significant investment, although many officers believe the costs at the outset are offset by the increased capabilities the LPR systems bring to their departments.[24]

The LPR systems' costs vary depending on vendors and camera configurations. Systems can be deployed on vehicles or portable trailers or attached at fixed sites. Cameras can cost $10,000 to $25,000 per unit for vehicle-mounted units, according to reports. Other configurations add surveillance capabilities or information-sharing with non-equipped vehicles; these cost more. Mobile LPR systems, which do not require extensive infrastructure to support, cost much less than fixed sites for the most part. Fixed sites can cost more because it is often necessary to work with other departments or utilities to place the cameras.[25]

Some fixed sites, such bridge mountings, can cost up to $100,000.[26] Some costs can be defrayed by attaching fixed units to existing infrastructure, decreasing site set-up costs and tapping existing lines to power LPR systems.[27] Though they are more expensive, fixed sites can scour geography of great interest to police or scan choke points like bridges or major intersections where parking a police cruiser could be difficult due to space or traffic concerns.

---

[24] "Automatic License Plate Reader Helps Jersey Police Fight Crime," *Homeland Security News Wire*, 22 March 2011.

[25] Roberts and Casanova, p. 21.

[26] Ozer, p. 30.

[27] Roberts and Casanova, p. 12.

System maintenance and data storage also can ring up bills for this technology.[28] If an agency's database can be made part of existing IT infrastructure, its maintenance costs can be lessened. Continuing advances in computer technology, processing speed, and data storage have slashed costs for system associated networks; data storage, especially for images, can be extensive and limiting. Cloud computing, stashing data on independent off-site servers, is a possible cost-cutter,[29] although this could create information security issues.

Survey data from George Mason University indicate that cost was the primary concern for police departments in putting in systems; 54 percent of agencies with LPR and almost 30 percent of agencies without it listed cost as their main concern.[30] Federal and state governments have invested substantial funding in Homeland Security grants to install these systems, especially for antiterrorism reasons.[31] In 2012, the U.S. Department of Homeland Security distributed more than $50 million in grants to law-enforcement agencies for these systems.[32] A $125,000 federal grant, for example, paid for cameras in Sumner County, Tennessee.[33] State agencies also paid for some of this equipment. In 2008, the Maryland Vehicle Theft Prevention Council bought seven cameras for Maryland counties.[34] These grants, which have supported LPR expansion nationwide, are subject to budget pressures. Further investment in this technology may be cut due to fiscal constraints at the federal and state levels. The George Mason University report, however, found that some departments turned to different revenue sources to finance their LPR systems: 10 percent of survey respondents indicated that they bought their equipment and maintain it with non-grant revenue, including asset-forfeiture funds, resources from private insurers, and routine agency budget sources.[35] This type of funding may become more common as grants disappear. Given the systems' benefits, many departments will strive to keep them running through tough budget times.

---

[28] The reports RAND examined did not contain any relevant data on the costs to police departments for LPR system maintenance and data storage. Several general assumptions can be surmised, however.

[29] Roberts and Casanova, p. 48.

[30] Lum et al., p. 25.

[31] Lum et al., p. 19.

[32] Margaret Rock, "The Era of License Plate Tracking," Mobiledia.com, 8 November 2012.

[33] "High-Tech License Plate Readers Aid Police but Raise Ethical Issues," *The Tennessean*, 6 May 2012.

[34] Matt Zapotosky, "Cruiser-Top Cameras Make Police Work a Snap," *Washington Post*, 2 August 2008.

[35] Lum et al., p. 19.

Benefits and Challenges

Benefits

LPR technology can be beneficial because it acts as a force multiplier for police departments in times when budgets and staffing are lean and uncertain. System cameras can scan thousands of license plates in an hour and automatically match them to hot-list databases, meaning that more license plates and therefore vehicles can be scrutinized and bigger areas patrolled effectively. With this technology, more arrests can be made because the LPR systems have detected suspect vehicles, leading to greater safety for both civilians and police. The technology can boost municipal revenue by identifying license plates associated with vehicles having outstanding citations or expired registrations.

The systems' main benefit is their scanning speed. Police previously had to manually enter plates into a mobile terminal or call the license number into a dispatcher to determine if a vehicle was stolen or sought for violations. Officers on patrol now are free to be on alert for other problems.[36] In Montgomery County, Maryland, a single officer with the system scanned more than 48,000 vehicles in 96-hour periods and across 27 days. The officer issued 255 traffic citations, identified 26 drivers with suspended licenses, caught 16 vehicle-emissions violators, found four stolen vehicles, and nabbed one expired plate.[37]

LPR started out as a way to combat auto theft, and nationwide anecdotal evidence indicates it is effective at this task. In New York City, arrests for grand larceny auto increased 31 percent in the first quarter of 2011 over the same period in 2010; since the city installed its network in 2006, it has recovered more than 3,600 stolen vehicles.[38] In Sacramento, California, the technology helped the city drop to number 13 from number 6 in per capita auto theft nationally; an officer said that the system cameras often identified multiple car thieves in an area.[39]

These LPR systems also can increase officer efficiency elsewhere. They can be used to determine what vehicles are in the vicinity of a crime scene, as previously mentioned, and their data can help to develop witness lists: Investigators can save time by not needing to spend long periods on time-consuming, door-to-door searches in neighborhoods seeking to learn who was in the area when a crime occurred.[40] One study found

---

[36] Jim McKay, "License Plate Recognition Systems Extend the Reach of Patrol Officers," *DigitalCommunities.com*, 8 April 2008.

[37] Roberts and Casanova, p. 23.

[38] Baker.

[39] Karen Wilkenson, "Technology Helps Probation Department Find More Stolen Vehicles," *Citrus Heights Patch*, 8 November 2012.

[40] "No License to Steal," *TECH-Beat*, National Law Enforcement and Corrections Technology Center, Gaithersburg, MD, Spring 2006.

that units in Cincinnati with this equipment were responsible for significantly more follow-up arrests than their counterparts lacking this tool.[41] During the study, LPR units were responsible for more than 70 follow-on arrests per month; traditional units conducted only 20 or so follow-on arrests per month.[42] If multiple agencies cooperate with the technology, allow universal access to data, and cover a whole region, LPR systems can retrieve stolen vehicles and home in on other suspect vehicles across jurisdictional bounds,[43] increasing the reach of law enforcement by providing information on violations outside the sway of individual departments.

LPR systems can be useful in identifying vehicles associated with unpaid fines, helping cities collect unpaid revenue. In the first twelve hours after New Haven's system went online, it identified 119 vehicles with parking violations. The city towed or disabled these vehicles with the "Denver boot," resulting in roughly $40,000 in fines for the city.[44] In New York City, through 2011, this technology was responsible for almost 35,000 summonses for parking violations.[45]

A last attribute of these systems is worth noting because so much has been written about police engaging in racial profiling. Some law-enforcement officers see plate reader technology as easing profiling concerns because cameras scan every vehicle within range irrespective of who is driving; one jurisdiction requires police to scan only vehicles' rear plate so its system does not photograph drivers. Although many departments scan front and rear plates, and some cameras photograph drivers, this occurs for all vehicles within camera range.[46]

Challenges

Challenges exist for law-enforcement agencies seeking to take full advantage of LPR capabilities. Its utility is directly tied to the amount and types of data available to the system. Technological issues, such as cameras inaccurately reading plates, can be problematic. Finally, how the units are deployed and who operates them can determine whether departments get maximum functionality for cost.

The LPR systems rely on timely data available to them for comparative readings. Databases can provide information on stolen vehicles, traffic offenses, and wanted and missing persons. Although all departments can access the Federal Bureau of Investigation's (FBI's) National Criminal Information Center (NCIC) vehicle theft databases

---

[41] Follow up arrests are defined as "arrestees who were not arrested during the incident but arrested after conducting follow-up investigations." Ozer, p. 49.

[42] Ozer, p. 74.

[43] "No License to Steal."

[44] Luke O'Brien, "License Plate Tracking for All," *Wired*, 25 July 2006.

[45] Baker.

[46] Police Executive Forum, p. 33.

and most can tap similar state databases, not all can access state warrant or DMV resources. The International Association of Chiefs of Policy (IACP) survey found only 60 percent of departments with LPR systems have access to warrant databases and 45 percent to Amber Alert data.[47] DMV data also are often unavailable. This reduces the technology's utility, limiting it to those crimes where relevant information is available for matching and comparison.

Databases also must be updated frequently for plate reader systems to work best. If a database is updated or downloaded to police LPR systems only once daily, reaction to recent crimes is restricted. If a vehicle is reported stolen the same day it is scanned by a system camera that received its updates the day before, no alert would be issued; the database would not record the vehicle as stolen. Although the NCIC database refreshes twice daily, survey data indicate that 43 percent of responding departments receive updates only once daily.[48]

Some technical flaws have also been identified in the technology. Although cameras usually are accurate, misreads occur; the systems' readers can have difficulty distinguishing between plates from different states. If a license number associated with a suspect plate happens to be identical in two states, the system may issue an alert for the plates from the wrong state. Vanity plates are especially susceptible to this problem.[49] Cameras also may struggle to identify plates that have been damaged, obscured, or coated with transparent materials.[50] They also have read by mistake the house address numbers on mailboxes[51] and characters on traffic signs. One database audit found the tag "ONEWAY" recorded almost 14,000 times.[52] Not only do these technical issues create the possibility of false alerts that could waste officer time, but storing these reads in a database could also complicate investigations. In the George Mason University survey, 23 percent of respondents listed technical concerns as a serious problem.[53]

System results also can be affected by how units are deployed: The Cincinnati police strategically put them in high-crime and gang-ridden areas and in other high-interest spots like parades, festivals, and driving under the influence (DUI) check-

---

[47] Roberts and Casanova, p. 27.

[48] Roberts and Casanova, p. 26.

[49] "No License to Steal."

[50] Bruce Taylor, Christopher Koper, and Daniel Woods, "Combating Vehicle Theft in Arizona: A Randomized Experiment with License Plate Recognition Technology" *Criminal Justice Review,* Vol. 1, Issue 27, Georgia State University, 2011, p. 3.

[51] Todd Corillo, "License Plate Reading Earns Waynesboro National Recognition," WHSV.com, Harrisonburg, VA, 29 June 2011.

[52] Julie Angwin and Jennifer Valentino-Devries, "New Tracking Frontier: Your License Plates," *Wall Street Journal Online*, 29 September 2012.

[53] Lum et al., p. 23.

points.[54] These tactics, as expected, can increase the number of alerts the LPR systems issue. But if they are deployed in haphazard ways, the technology's utility drops; only 24 percent of responding agencies in one survey had created policies on deployment.[55]

While it appears these systems increase the recovery of stolen vehicles, it is unclear if they curb vehicle thefts. The 2010 study by George Mason University and a 2011 study by the Police Executive Research Forum sought to quantify the reduction in auto thefts attributable to LPR systems in Northern Virginia and Arizona, respectively. Both studies determined that the systems' benefits were inconclusive in this area. Both studies concluded that the systems scan more plates than manual searches and thus identify more stolen vehicles; the incidence of auto theft, however, did not decline after the tests were concluded.[56] It could not be determined whether expanding the technology, apprehending more car thieves, or criminals' greater knowledge of the LPR systems' existence would result in fewer stolen vehicles. More research here would be useful.

Privacy Concerns

The rapid adoption of LPR by agencies across the United States has led to concerns over how its data are used and who has access to this information. License plate reader systems provoke privacy advocates generally for reasons ranging from "Big Brother" concerns about governments' watching and tracking citizens to specific instances focusing on how information is used and procedures governing database storage and access.[57] These databases can retain information for long periods and detail vehicle locations, thus potentially disclosing information about vehicle owners or other drivers. One report argues that recording information that could be associated with motorists' driving habits could raise First Amendment concerns because license plate readers may place motorists' license plates at such places as addiction counseling centers or staging areas for political protests.[58] Fears about police abuse of this information have led civil rights groups to demand access to official procedures and data and some legislative efforts to ban this technology's use.

The technology also raises concerns focusing on whether its stored data are "personally identifiable information (PII)," an especially sensitive category of individuals' information that can help distinguish individuals and may be used to track an individ-

[54] Ozer, p. 31.

[55] Roberts and Casanova, p. 26.

[56] For more information, see Lum et al. and Taylor et al.

[57] Lum et al., p. 10.

[58] Angwin and Valentino-Devries.

ual's movement.[59] Many law-enforcement officers argue that the databases record only vehicle information[60] and, further, that because cameras reside in public places, individuals should hold no expectation of privacy. The license plates are a "gateway" to PII, and although the LPR system records the plate information, it is the investigative process that identifies individuals.[61] As part of the 1994 Driver's Privacy Protection Act,[62] DMVs are prohibited from disclosing personal information related to motor vehicles for anything other than official use.[63] Therefore, many law-enforcement officials see a wall between the license plate number and the PII associated with it.

Privacy advocates counter that vehicle information can provide intimate details of people's lives; they say that it violates individuals' rights to keep records on vehicles not suspected and not involved in criminal activity.[64] This "chilling" effect can make some people more cautious of exercising their constitutional rights such as attending political rallies.[65] Since this technology spread so quickly and so many agencies lack policies and laws governing its use, as well as on its access and its data retention, an atmosphere has been created in which abuse is possible.[66] This hardly has been helped by the absence of a single set of regulations on use and access to data for these systems; many departments not only have no set policies, they did not research legal concerns before acquiring the technology. Only 28 percent of those responding in the George Mason University survey said they had investigated legal concerns of this technology before adopting it.[67] And only 17 percent of responding departments stated that legal or privacy concerns were a factor considered in their system's use.[68]

In 2013, the American Civil Liberties Union (ACLU) published a report outlining privacy concerns about this technology after sending public records requests to nearly 600 law-enforcement agencies and receiving more than 26,000 pages of documents on police procedures in this area.[69] The ACLU concluded that long data retention periods and more information-sharing among law enforcement could have negative effects on personal privacy, stating:

---

[59] Roberts and Casanova, p. 30.

[60] Roper, "Cops Move to Protect License Plate Data."

[61] IACP, *Privacy Impact Assessment Report for the Utilization of License Plate Readers*, Alexandria, VA, 2009, p. 9.

[62] Drivers Privacy Protection Act, 18 U.S.C. sec. 2721 et seq.

[63] IACP, p. 10.

[64] Ben Eisler, "ACLU Concerned Automatic License Plate Readers May Invade Privacy," WJLA.com, 30 July 2012.

[65] IACP, p. 13.

[66] Angwin and Valentino-Devries.

[67] Lum et al., p. 60.

[68] Lum et al., p. 25.

[69] ACLU, p. 3.

> Longer retention periods and more widespread sharing allow law enforcement agents to assemble the individual puzzle pieces of where we have been over time into a single, high-resolution image of our lives. This constant monitoring and permanent recording violates our privacy in a number of respects.[70]

The ACLU believes that the sensitive personal information associated with license plates stored in these system databases could lead to abusive tracking by individual officers, institutional abuses, and discriminatory targeting. It decried police tactics, such as driving LPR-equipped police cars around mosques and recording license plates of those inside.[71] The ACLU believes that widespread use of similar tactics might dampen the free exercise of many constitutionally guaranteed rights.

Two of the more critical aspects of the privacy concerns regarding LPR data are the length of time LPR records are retained and who has access to the LPR database. Other factors, which we will discuss in Chapter Five, such as who owns the data, do play an important role, but retention and access appear to be the most crucial. It is important to note that privacy concerns regarding LPR data are not necessarily case use–specific. Privacy concerns center mainly on reads kept in the LPR database that were not associated with activity of interest to the police when the reads occurred. These records can be exploited at a later data at the discretion of the police. Such records are generally associated with analytic case uses, making reactive case uses less of a privacy concern. As the ACLU states, "License plate readers would pose few civil liberties risks if they only checked plates against hotlists and these hotlists were implemented soundly."[72]

Departments with short data retention, whether by policy or due to technical storage limits, are less subject to privacy concerns compared with agencies that keep data for longer periods. Data that are kept for short periods are less susceptible to abuse because it is difficult to gain personal information from the patterns and travel routes of a license plate stored in LPR databases. Likewise, departments that restrict access or require need-to-know authorization before past records can be examined are less susceptible to abuses. Departments that provide stored LPR data to any personnel for any reason make privacy advocates question the effectiveness of preventing the information being used for nefarious purposes.

Figure 1.1 in Chapter One illustrates the usefulness of LPR systems based on data retention and access to external databases. This taxonomy can be applied to privacy concerns. As noted above, data retention policies are an important privacy concern. Access to external databases is replaced by the amount of access is available to the LPR database containing the camera reads. Greater access, to the public or a large number

---

[70] ACLU, p. 7.

[71] ACLU, p. 11.

[72] ACLU, p. 2.

of departmental personnel, creates more privacy risks because the greater number of people who have access to the information increases the possibility of abuse. In contrast, departments that restrict access or require need-to-know authorization decrease privacy risks. Figure 2.1, which uses the same taxonomy, illustrates the dynamic between LPR privacy risks and retention and access polices.

Policies on data retention vary widely among jurisdictions. Data for LPR systems, in theory, can be kept indefinitely. Lower costs and higher capacity for storage may encourage departments to keep data for as long as possible; some reports indicate that data are retained for up to four years after an initial alert and may be kept indefinitely.[73] For policy or space considerations, it appears that many departments keep data for much less time. Maine state law requires data purges every 21 days, while regulations in Columbia, Missouri, require the information be deleted after 30 days.[74] While many other departments have similar requirements, Minneapolis, Minnesota, conversely keeps data for a year and makes it publicly available.[75] A number of departments lack policies on data retention; in the recent IACP report, only 48 percent of responding departments stated that they had created such policies.[76]

Figure 2.1
Privacy Risk for LPR

Retention policies	Database access by personnel: Least access	Database access by personnel: Most access
Longest retention	Moderate risk • Retains data for extended periods • Restricts or requires authorization for data retrieval	Most risk • Retains data for extended periods • Allows significant access to personnel
Shortest retention	Least risk • Purges data in short order • Restricts or requires authorization for data retrieval	Moderate risk • Purges data in short order • Allows significant access to personnel

RAND *RR467-2.1*

[73] Mark Davis, "Vermont Bill Targets License Plate Readers," *Valley News*, West Lebanon, NH, 23 January 2013.

[74] "High-Tech License Plate Readers Aid Police."

[75] Eric Roper, "Man Uses License Plate Data to Repossess Car in Mpls," *Minneapolis Star Tribune*, 30 August 2012.

[76] Roberts and Casanova, Executive Summary.

Questions have also been raised about who has access to these records. Retaining them over long periods can provide information on individuals' movements and lifestyles, and some privacy advocates and legislators argue that access to these data without probable cause can lead to law-enforcement abuse.[77] As one California legislator stated recently, "Should a cop who thinks you're cute have access to your daily movements for the past ten years without your knowledge or consent?"[78] This legislator recently introduced a bill in the California State Senate to limit access to these databases only to officers armed with warrants for information from them.

Access to these data also varies among jurisdictions. Some agencies allow only a handful of officers with additional training in privacy issues to access information in databases. Concerns arise when unauthorized personnel access data or the information is used for other-than-official purposes. Although many agencies (68 percent) have created policies on access to system data,[79] only 6 percent of agencies identified misuse or hacking of this information as a concern, according to recent reports.[80]

More concerns arise when the data are public. In Minnesota, the database is considered a public record and is available upon request. As of late 2012, authorities have received 80 or so such requests, many for legitimate purposes, such as by those legitimately repossessing vehicles or bounty hunters. Law-enforcement officials themselves have argued that LPR information could be used by criminals to target unsuspecting citizens.[81] As Minneapolis Deputy Police Chief Robert Allen told the Minneapolis Star Tribune, "If, for example, a stalker wants to see where their prey has been, they can do public records search and we are required to provide them with information about where that vehicle has been seen by our system."[82] Some jurisdictions have banned or refused LPR equipment. Norwich, a small town in eastern Vermont, recently turned down a grant to purchase system cameras, citing privacy concerns.[83] The attorney general of Virginia, citing privacy concerns, issued a recent opinion stating that system cameras may be used only "when there is a criminal investigation or an immediate threat to public safety."[84] The opinion explains that "[i]nformation shall not be collected unless the need for it has been clearly established in advance." Law enforcement may not simply create a database of vehicle locations, the vast majority of which have nothing to do with any criminal violations. Legislation recently introduced in South

---

[77] Davis.

[78] Rock.

[79] Roberts and Casanova, p. 25.

[80] Lum et al., p. 25.

[81] Roper, "Cops Move to Protect License Plate Data."

[82] Roper, "Cops Move to Protect License Plate Data."

[83] Davis.

[84] Allison Klein, "Cuccinelli Limits License-Plate Cameras," *Washington Post*, 8 March 2013.

Carolina would ban the technology's use statewide. Citing questions about what the government could do with LPR data and how it would be protected, the introduced legislation would bar LPR systems until privacy and access issues can be resolved.[85]

Others fret about private companies putting up plate readers to acquire similarly large amounts of data about individuals; according to anecdotal information, private companies have focused up to now on repossessing automobiles from those who default on loans. But the technology's capacity to collect volumes of data may interest marketers specializing in consumer behavior. While plate data typically are not linked to individuals, many companies, such as car dealerships or parking garages, may possess the information to do so. Government agencies also could then subpoena that private information and link it to individuals.

Clearly, when a technology spreads so fast and carries with it elements that create serious privacy concerns unfettered by carefully considered policy or long practice, analytic research on its opportunities and obstacles is warranted. This study seeks to fill this need and its next chapter describes the RAND methodology that informed our research strategy. It can be argued that privacy concerns are the most significant challenge to LPR use. This being the case, a more detailed and extensive examination of the legal aspects of privacy concerns regarding LPR use is provided in Chapter Five.

---

[85] Robert Kittle, "Bill Would Ban Use of Automatic License Plate Readers in SC," WSPA.com, 2 January 2013.

CHAPTER THREE

Methodology

There is a dearth of academic research on the benefits and challenges of the rapidly spreading LPR technology. Existing studies have provided aggregated survey results, are advocacy focused, or have targeted a single-use case. Earlier works only touch on peripheral issues, such as interoperability, system structure, overall system cost, and operational procedures across a broad spectrum of uses. In developing our methodology, we sought to perform nuanced and detailed research, gathering the perspectives of a variety of personnel involved in installing, using, and maintaining LPR systems.

To gather needed information, RAND carried out case studies through exploratory interviews with personnel in various types of law-enforcement agencies. These interviews targeted not only users but also administrators, analysts, budget experts, and senior departmental leadership. This approach seeks to expand the knowledge base about LPR systems by providing rich, contextual information from those most knowledgeable now about the weaknesses and strengths or incentives and barriers to this technology's effective implementation and use. To select participating departments, RAND developed an analytic framework to identify those with unique environments for their LPR systems, thereby ensuring a diverse and interesting representation. RAND then scrutinized relevant demographic and crime-rate statistics to select specific departments. We took care to ensure that each end of the statistical spectrum for each selection criteria was represented, as is described below.

Study Population Categorization Scheme

RAND conducted seven case studies.[1] The unit of analysis for the case studies was an agency. The universe for case study selection was all law-enforcement agencies in the United States, and the study population was agencies in the 2007 LEMAS survey

[1] RAND originally intended to conduct eight case studies, with the eighth case study being a department that had previously used the technology but discontinued it. The research team felt this type of department would provide more information on system challenges. The research team could only find one such department, which declined to participate.

database. That 2007 survey specifically asked participating departments if they used LPR. The RAND team conducted additional research, beyond the survey, to identify 173 departments employing the technology. This list formed the basis of the sample set for this study. The survey also provided the number of sworn officers for each department in our sample set, from which we selected agencies to fit the department size and population category, as described below.

To collect as much unique and relevant data as possible, RAND organized the sample set of departments into three categories designed to address diversity in system users nationwide. We then selected case study candidates from within each category. The categories were (1) department size and population, (2) border agencies, and (3) geographic clusters. Our selections aimed to capture agencies of different sizes, those operating on the border, and some operating in a regional cluster. Our assumption was that agencies from each category would bring a unique view about the technology and could share information on its benefits and challenges distinctive to each department.

Department Size and Population

We looked at department size and a given jurisdiction's population so we could examine differences in "Big v. Small" departments. RAND's hypothesis was that each would use the technology differently, finding its advantages and challenges depending on the agency's size and available resources, i.e., budgets, staffing, etc. Larger departments were assumed to have larger LPR systems, with more analytic support, which could be applied in more case uses than smaller departments. This category also could be used to examine differences in data access: Larger departments also could be assumed to have more personnel with access to system data, necessitating clear regulatory policies. Smaller departments could be assumed to have less infrastructure and maintenance requirements, affecting how they would use their LPR systems.

Border Locations

We also examined how this technology is used on both the northern and southern U.S. borders. These departments operate in unique environments with high international vehicle traffic from Canada and Mexico; we thought that examining issues about this traffic was critical to understanding the utility of plate recognition systems. Southern border locations could be expected to home in on vehicles related to crimes committed by Mexican drug cartels and in immigration matters, while northern border locations were deemed likely to use the technology for activity related to tourism and standard U.S.-Canadian border commerce. By selecting international border locations, RAND also hoped to generate information to inform policy questions about cooperation and information-sharing among U.S. federal border and national security agencies, as well as America's international partners.

Geographic Clusters

Our last category deals with multiple agencies in the same geographic areas. Ideally, RAND would gather information from them on the interoperability among systems of different departments and on the levels of cooperation and information-sharing occurring with LPR use across jurisdictions. Because different departments could be assumed to have different regulations and policies on the technology's use, data access, and data retention, the interplay among departments in a regional setting would provide useful information on the benefits and difficulties of sharing information across jurisdictional boundaries.

Selection Criteria Within Categories

To select agencies for the case studies, we tried to choose agencies spanning a range in several characteristics, including size and per capita income of the population served, number of sworn officers, violent crime rate, overall property crime rates, and vehicle theft rates. RAND identified departments with LPR systems on the northern and southern borders. We considered how recently departments had adopted plate readers so that we could examine agencies that had had LPR systems for many years and could be assumed to have extensive operational and administrative experience and those that were in the initial stages of use so they were grappling with start-up challenges. Although our subsequent discussion of findings does not distinguish agencies according to several of these characteristics, choosing agencies that spanned these characteristics helped increase the representativeness of our findings.

For population and per capita income information, RAND turned to data from the 2010 U.S. Census. Population size affects the operations of each department. It could be assumed that the LPR systems in more populous areas would provide more alerts and be used more extensively. When considering resource levels, we did not have access to comprehensive budget data for all 173 departments in the sample. Instead, we used Census per capita income data for each jurisdiction. Specific budget data for each department would have been ideal to determine the funding support for LPR systems. Lacking such data, per capita income at least would show the tax base available to each department.

The last characteristic was crime rates. Since research into LPR technology is not yet as mature as other aspects of law-enforcement activities, RAND chose two general statistics, violent and property crime rates, and one specific statistic, vehicle theft rates. Vehicle theft rate was an obvious choice considering historical use of LPR in combating vehicle theft. Violent and property crime rates informed the general crime environment of the departments and the frequency of opportunity for LPR to contribute to more extensive criminal investigations.

After statistics were compiled for all 173 departments, the sample set was subdivided into agencies that fell both above and below the median levels on selection characteristics for each of the three, main sample categories. Within these divisions, specific agencies were selected using our own subject matter expert consultations and logistical considerations. Table 3.1 identifies the selected departments for this study and provides the relevant statistics and other information for each department. DMV-Enabled Cluster County, Cluster City, and Cluster County are three departments in and around a major metropolitan area that were selected to study interoperability issues. DMV-Enabled Cluster County is a county-level jurisdiction with access to data from the DMV. Cluster County is also a county-level jurisdiction bordering both DMV-Enabled Cluster County and Cluster City. Cluster City is a city-level jurisdiction that borders both DMV-Enabled Cluster County and Cluster County. Small Town is a small department in a rural area with a population well below median. Big City is a larger department in an urban area with substantial population, both within city limits and in neighboring suburban areas. Summaries of each case study are provided in Appendix A.

Interview Protocol

Once the departments had been selected, RAND developed a comprehensive protocol for conducting interviews so we could gather information on all relevant topics matching our study objectives. We also wanted to let our interview subjects address issues not part of the initial protocol, giving more in-depth information and adding to the knowledge base about this technology. This section identifies key protocol topics

Table 3.1
Statistical Criteria for Selected Departments

Department	Population	Sworn Full-Time Employees	Per Capita Income ($)	Violent Crime Rate[a]	Property Crime Rate[a]	Vehicle Theft Rate[a]	Recent Adopter	Northern Border	Southern Border
Median	201,165	267	26,100	307.5	2,730.7	152.5			
DMV-Enabled Cluster County	A	A	A	B	B	M	N	N	N
Cluster City	A	A	A	A	A	A	N	N	N
Cluster County	M	A	A	B	B	B	Y	N	N
Small Town	B	B	B	A	A	M	N	N	N
Big City	A	A	B	A	A	A	Y	N	N
South City	A	A	B	A	M	A	N	N	Y
North Town	B	B	B	A	A	A	N	Y	N

[a] Per 100,000.

NOTE: A = above median, B = below median, M = at median, Y = yes, N = no.

and gives the rationale for including each. A complete copy of the protocol appears in Appendix B. Table 3.2 presents the major study topics.

After the protocol was developed, RAND contacted law-enforcement personnel from each department, identifying those we assumed to have the required knowledge of the agency's deployed LPR systems and seeking their participation. Interviews were conducted on-site at each department between September 2012 and December 2012, and we talked to field users, system administrators, finance officials, analysts, and senior departmental leadership. We got several field demonstrations of the equipment and rode with police on patrol as they employed their plate reader systems. We saw firsthand how the equipment operates and the level of detail for images it returned. We saw how alerts popped up for officers on mobile terminals. We inspected trailer units at one department. This greatly enhanced RAND's understanding of LPR use in the field. RAND also inspected trailer units at one department.

The level of access granted and the information provided by departments for the case studies fulfilled RAND's expectations for an effective qualitative analysis of

Table 3.2
Case Study Topics and Rationales

Topic	Rationale for Inclusion
Major case uses	Determine if auto theft is still the most prevalent use for LPR Determine what other types of cases benefit from LPR Determine how much LPR data used analytically to support investigation
Interoperability/ information sharing	Determine if independent systems can pass data between them Determine the level of information sharing between departments Determine if standardization of LPR records exists
Cost	Identify initial funding mechanisms for LPR deployment Identify operational and maintenance costs
Implementation/ system structure	Determine ratio of fixed v. mobile LPR cameras Explain initial rationale for department acquisition Identify how the system is administered and who has data access Identify technical and data storage issues
Privacy	Identify privacy and legal concerns Examine departmental policies regarding data retention and access Identify level of legal advice provided to department on privacy issues
Operational policies	Identify how field officers respond to LPR alerts Identify evidentiary procedures associated with LPR records Identify types of databases department has access to
Best practices/ lessons learned	Determine what policies and procedures work best in regards to LPR use

LPR use. RAND gathered information on all the study topics and uncovered valuable themes concerning this technology and its uses, as we describe in Chapter Six. These themes, we hope, will enhance the understanding of issues relating to systems' deployment that will be useful for those making decisions to buy and use them. The next chapter compares the case study departments, using the site selection framework to highlight similarities and differences in how officials use this technology.

CHAPTER FOUR

LPR Uses in Different Operational Environments

We used three analytical lenses that would provide unique perspectives on how LPR systems are used and structured. We concentrated on relevant elements of the technology where we determined we could add to the knowledge base of literature on this topic. We focused on categories dealing with resources, information-sharing, operating environment and policies on these systems' use.

First, we examined this technology's use based on department size and jurisdictional population. This let RAND see how differences in resource levels and sizes of operating environments affect police use of plate reader equipment. We also explored the unique environments for LPR systems' use in jurisdictions with international borders. Assessing the systems' utility in such environments can give insight into the technology's flexibility and effectiveness on nontraditional case-uses. We turned to a third analytical lens to look at departments at various jurisdictional levels and in the same geographic area. By examining clustered agencies we hoped to discern the level of information-sharing among system-equipped departments in the same area; we wanted to understand individual LPR systems' interoperability. Because we may assume that different departments have different policies and procedures on LPR use in the field, as well as on data retention and future investigative use, understanding how these affect information-sharing was a key aspect of assessing the systems' overall utility.

Access to Resources: Big v. Small

To examine the relationship between department size and LPR use, RAND selected a small department in a rural area with a population well below the median[1] and a larger department in an urban area with substantial population, both within city limits and in neighboring suburban areas. We chose Small Town and Big City (see Table 2.2 of the previous chapter). Small Town has a population of 40,000 or so and a sworn, full-time police force of slightly more than 100. It falls below the median per capita income level, has a slightly higher-than-median violent crime rate and a higher-than-median

[1] In this context, "Below Median Population" refers to the median of departments in RAND selection sample.

property crime rate. Its rates of vehicles theft are at median levels. In contrast, Big City is larger, with a population of roughly 300,000, plus more residents in suburban areas outside city limits. Its police department has more than 1,000 full-time sworn employees. Per capita income of its residents is slightly below the median, and the city has rates of violent crime, property crime and vehicle theft well above median levels for the selection sample. Big City also is involved in a regional law-enforcement cooperative whose participating departments often share information about many different kinds of law-enforcement activities, including plate reader data.

Small Town employs a system with two vehicle-mounted cameras, one front-facing and one rear-facing on two patrol vehicles, to identify stolen vehicles while officers patrol; the system sends alerts corresponding to terrorist watch lists. Small Town has access to the FBI's NCIC hotlist and its own state hotlists; there is no local equivalent that identifies local warrants and suspect vehicles. Because its home state does not provide DMV information to individual departments, Small Town's system is not used for traffic-related violations, expired driver licenses or plates, emission violations, or insurance lapses. Police from the town say they get only a handful of alerts each month and they pursue all of them.

The license plate reader units have helped to identify vehicles in the vicinity of a crime scene. The technology sometimes is used for querying stored records for specific investigations; this is restricted by personnel and budget limits. The system cannot share information electronically with any other agencies and any data requests must be made manually. Small Town uses its system in a narrowly focused way for only a few case-uses. The lack of information-sharing with its state DMV inhibits the town from taking more advantage of this technology. Personnel restrictions limit the data that can be analyzed from the system database for investigative purposes, and the inability to share data electronically with other departments prevents Small Town from taking advantage of other police LPR systems. This would be expected from a small, rural department. Though useful for specific purposes, LPR technology is unlikely to be as effective a force-multiplier as it may be elsewhere. Although officers in camera-equipped cars may see their workloads reduced by eliminating the need to manually input licenses when searching for violators, this advantage shrinks because so few police cars carry the technology.

In contrast, Big City has an extensive, multi-layered license plate reader network that shares information among multiple agencies and on a diverse set of investigations. It employs more than 100 mobile and at least four fixed cameras and can access several dozen more mobile and fixed cameras from suburban and county police in its area. Big City participates in an integrated, regional law-enforcement cooperative that funds various crime-fighting initiatives, and it has been a principal driver in expanding plate readers in the region.

Big City uses system information for both rapid response and retrieving stored records. The large number of cameras and the interconnectivity of departments in the area allow for rapid information-sharing among jurisdictions, facilitating investigations that cross jurisdictional boundaries. If a system camera in one jurisdiction spots a stolen vehicle, that information is automatically available to participating departments; this means that vehicles can be tracked in broader areas, leading to more recoveries. Officers use discretion to determine if they should act on a system alert. This lets it record the locations and times of minor violations but also to serve up information directly to officers on more serious crimes.

Big City operates a fusion center that incorporates license plate reader data from more than 80 regional law-enforcement departments and employs several full-time analysts who maintain access to an investigative database. Although Big City does not use DMV-related information with its system, the volume of records available to it, more than 20 million per year, allows it to apply data in any type of investigation. It receives both the NCIC hotlist and its state's hotlists daily; it has access to a locally generated county list of local vehicles of interest; and it gets state and national alerts. Departments in its area that do not have license plate readers still can access data in the system database via analysts in the fusion center.

Thus, this technology truly acts as a force-multiplier for area law enforcement. The fusion center helps investigators detect vehicles in the vicinity of crime scenes, corroborate alibis, and identify partial plates of vehicles given to police by witnesses. The large number of cameras allows more officers to spend less time manually inputting license plates while looking for violations, increasing police efficiency on patrols. The information-sharing builds more cooperation among departments in prosecuting crimes that cross jurisdictional borders.

The evidence just cited points to LPR being more useful for larger departments with more resources, more personnel, and the ability to connect neighboring departments and to share information. Although this technology has utility for smaller departments, the environment in which they operate may diminish the LPR systems' distinct ability to lessen workloads and share information with others so as to see full benefit from this technology. The inability to share information within a department or to receive data from an outside agency prevents the increased levels of cooperation evident in larger areas that cooperate regionally. Further analytic study would be useful here to determine whether the RAND observations, made from in-depth information gathered about two departments, accurately represents divergences between big and small departments nationwide. The technology's benefits for small departments may be more apparent where there are robust structures to support regional or statewide information-sharing.

LPR on the Border

Police on international borders operate in distinctive environments, dealing with often unique matters related to tourism, international commerce, and cross-border criminal activity. These border areas also seem ideal spots to examine issues about license plate readers that may differ from those in the rest of the country. The RAND research team looked at two departments, one on the northern border with Canada and one on the southern border with Mexico, to study international traffic and cooperation with federal and foreign law-enforcement agencies.

We chose two departments, North Town and South City, to examine how LPR works on the border. North Town has roughly 50,000 residents and a large tourism industry from both the United States and Canada. While it is in an area popular for both out-of-state and foreign visitors, North Town's police force is relatively small, with 150 sworn, full-time employees. The city has a below-median per capita income. North Town is above median in crime rates in all the categories RAND examined. On the southern border, South City is much larger, with 650,000 residents and a police force with more than 1,000 sworn full-time employees. It sees a substantial amount of cross-border movement of Mexican and other foreign nationals into and out of the state. It also has a significant amount of both legal commerce and illegal criminal activity across the border. South City deals with median levels of violent and property crime and above-median levels of vehicle theft. But it is important to note that anecdotal information from South City authorities indicated that many violent crimes are associated with activities by Mexican drug cartels. These crimes are not reported to police, thereby, artificially decreasing the official statistics.

North Town has a small plate reader system with only three cameras. Two cameras are mounted on patrol vehicles; there is a third mobile unit. Five more mobile units are on order. The system was first obtained under Homeland Security grants to cover multiple industrial sites in the area. Although the region has international traffic from Canada, criminal activity associated with this flow is low, so North Town uses its system in more-traditional ways. It is connected to its state's DMV databases, so the system's most important targets are suspended registrations and lapsed insurance and parking infractions; a smaller portion of its alerts is for recovery of stolen vehicles.

The interaction of LPR systems is low among North Town, U.S. federal authorities, and Canadian officials. Both U.S and Canadian border enforcement agencies maintain such systems at border crossings for their respective countries. The U.S. system is not integrated with local LPR databases, and cooperation on vehicle investigations between North Town and U.S. federal authorities occurs only on an ad hoc basis. North Town law enforcement says this cooperation is uncommon. It works in similar fashion with Canadian authorities—ad hoc, limited, and informal. Diplomatic requests must be made, which slows the process to gain access to Canadian data for U.S. investigations; thus, these procedures normally are time-prohibitive for investigations. Some information passes informally but infrequently and this practice is based in personal relationships.

South City operates a larger system, with two dozen cameras mounted on vehicles and trailers and with several portable units. These units were purchased with state border-security funding, targeted for tracking vehicles exiting the United States, and South City, unlike North Town, primarily uses its system to recover stolen vehicles. While vehicle theft is a common problem nationwide, South City's situation permits plate readers to play a unique role in helping police. That is because its location exposes its area to substantial narcotics-related criminal activities by drug cartels. Narcotics trafficking, murder, property crime, and other drug-related violence are common.

For example, the cartels typically use stolen vehicles to move drugs and people across borders. Because a number of stolen vehicles are tied to drug activity, plate readers provide South City with the opportunity not only to recover these cars and trucks but also to uncover clues to other crimes. South City officers said that when they recover stolen vehicles, it often leads them to discover forensic and other evidence linked with other crimes, providing leads on drug-related murders, drug-trafficking routes, and smugglers and kidnappings. The technology, thus, can act as a significant force-multiplier, providing value not only in the recovery of stolen vehicles but also the prosecution of serious and major drug-related crimes.

South City sees insignificant interaction with U.S. federal or international law enforcement over LPR data (just as occurs in North Town). South City employs the technology to cover the border, identifying suspicious vehicles exiting the United States into Mexico. The U.S. Border Patrol maintains its system cameras for inbound vehicles. South City lacks access to Border Patrol system databases, but these data are provided to the National Insurance Crime Bureau (NICB) and city authorities can access the information there, meaning that information from in-bound Border Patrol cameras is available to them. There does not appear to be any direct connection between federal and South City databases for investigative purposes. Cooperation exists on an ad hoc basis; no formal information-sharing has been established due to concerns over the flow and quantity of data shared.

LPR information-sharing with Mexico is virtually nonexistent. South City has one officer who maintains contact on border issues with his Mexican counterpart, but intelligence-sharing is minimal. There appears to be concern in South City, and among other U.S. agencies, that Mexico cannot protect U.S. data and that Mexican data would be corrupted and unreliable. There is no official cooperation for LPR data between South City and Mexico.

Examining LPR in border environments revealed that although LPR can identify vehicles entering and exiting the United States, there is little to no official sharing of LPR data between federal agencies and local police departments. This degrades LPR usefulness because information does not flow between law-enforcement agencies and robust cooperation on issues that LPR benefits is minimal. The border environment for

South City, though, does provide ample opportunities for LPR to contribute greatly to law-enforcement efforts. As stated previously, LPR's effectiveness at recovering stolen vehicles, even in a high-crime and transient area such as that of South City, provides the agency with additional information on related crimes and enhances the police's ability to investigate those crimes. An area with substantial tourism, such as North Town, does not provide the same opportunities, diminishing LPR usefulness. However, increased information-sharing with federal authorities could increase LPR's ability to aid police in high-tourism areas.

Geographic Clusters

By examining police agencies operating in the same geographic region, RAND obtained insights on interoperability of LPR systems; information-sharing among departments; and how different agencies create policies and procedures governing this technology's use and information. In theory, departments clustered nearby would be more likely to share experiences and operational information and to coordinate activities. This cooperation and interoperability, as has already been established in examinations of resource levels and operating environments, could increase the utility of LPR systems by standardizing policies and providing easy ways for agencies to share information. RAND chose three departments in and around a major metropolitan area to consider this possibility: (1) DMV-Enabled Cluster County, (2) Cluster City, and (3) Cluster County.

DMV-Enabled Cluster County is a county-level jurisdiction with access to data from the DMV. With a population of roughly 1 million, DMV-Enabled Cluster County borders both Cluster City and Cluster County and has a significantly higher per capita income than the sample median. DMV-Enabled Cluster County's police force numbers 1,200 or so sworn full-time employees; it is in an area with crime rates generally below median, leaving it relatively wealthy and in low-crime area. Cluster County is also a county-level jurisdiction bordering both DMV-Enabled Cluster County, and Cluster City. It has a population of 200,000 with a police force of slightly less than 400. Per capita income for the county is more than double the sample median with crime rates below or at median, which—like its big neighbor—makes it relatively wealthy and in a lower crime area. Cluster City borders both DMV-Enabled Cluster County and Cluster County. It has 600,000 or so residents, a police force of approximately 4,000 sworn, full-time employees, and an above-average per capita income. But its crime rates are two to four times above the sample median: Cluster City has substantial areas of high crime.

The three departments and several other neighboring jurisdictions bought LPR technology with grants awarded to a local area law-enforcement governing council. It allocated funding for cameras by department size to each participating departments;

the systems were ostensibly a counterterrorism tool to identify threats traversing the region. Although the departments had the same funder, there was no mechanism initially in place for them to share information, and this still inhibits cooperation.

Though the departments could choose their own vendors, they used the same one, a chance to address interoperability that otherwise may not have been possible. But just because they have the same LPR systems, which theoretically would allow cooperative sharing of information on alerts, little of this has occurred; bureaucracy has hindered mechanisms for sharing.

Both DMV-Enabled Cluster County and Cluster City say that to share information—with outsiders and electronically—they must put in place individual memoranda of understanding (MoUs). This take time and the MoU must grind through a bureaucracy. Thus information is shared only in manual fashion with officers contacting neighboring jurisdictions to seek information on specific license plates. This is a slow process because agencies must assign staff to respond to those requests.

DMV-Enabled Cluster County has access to local DMV data—and is the only jurisdiction in the cluster that has it. Cluster City does not share data electronically with any other agency and it must make manual requests to receive plate information from outside departments. Cluster County is more effective in sharing information: It can electronically connect its license plate reader system data to several nearby agencies but not to the DMV. This real-time passing of information cuts the time needed for officers outside Cluster County to retrieve this information, increasing officer efficiency and allowing for quicker response times to alerts.

Examining policies of police departments in the same area is also an indicator of how departments view LPR's utility and associated challenges. One of the most important policies regarding LPR employment is retention of LPR data. Many privacy concerns have been voiced about a police department's ability to track individual citizens using stored data, and a department's retention policy reflects its concerns in this area. Examining this particular cluster of departments reveals substantial differences in the length of time LPR data is accessible and the reasons why the information is retained. The differing lengths of data storage time could raise an interesting set of privacy concerns specific to a cluster of departments that shares LPR data.

Agencies in the cluster keep the data anywhere from six to 24 months. Cluster County keeps all scans for six months and alert information for 24 months. Cluster County discards non-alert reads after six months, due to concerns both about privacy and the manpower required to respond to Freedom of Information Act requests for the data. Alert reads are kept longer to support investigations. DMV-Enabled Cluster County keeps all data for only 12 months because of privacy concerns; it has made provisions to keep certain investigative reads indefinitely—those that may be applicable to cold-case investigations and thus may be needed for long periods. Specific policies and accesses were created to protect this information to ensure confidentiality and privacy. An officer must make and explain an access request, which can only be granted by the

DMV-Enabled Cluster County's police chief or assistant chief for specific cold-case investigations. Cluster City keeps all data the longest: 24 months for all reads. It does so, it says, because that is its technically feasible maximum and because it says it seeks to adhere to federal regulations on retention of personally identifiable information. Cluster City interprets federal regulations as permitting retention of these data for 36 months, which, due to storage limits, is technically unfeasible.

Although each agency may have its own rationale and they all share many commonalities (they are neighbors with a single grant funder, they cooperate on other law-enforcement issues, and they use the same systems), the diverse policy approaches the clustered agencies take may cause problems in the future. Privacy advocates may press the departments to standardize their policies. If they resolve their blocks to data sharing, it still is conceivable that the department with the longest retention policy could share personally identifiable information with other departments; this would negate the push to discard plate information quickly due to privacy concerns. Privacy advocates conceivably could still argue that even if one agency deletes its own data, it still could track an individual with information shared by other departments retaining information longer. This could impede sharing among departments, reducing their cooperation and limiting the very utility of plate systems.

Many things affect the utility of LPR technology. Operational environment, resources, and departmental cooperation and policies all impact how LPR is used and contribute to potential challenges. Evidence suggests that, as may be expected, LPR technology is more useful to departments with more resources that operate in higher crime areas and readily share and receive information with other agencies and departments. Nonstandardized policies, lack of access to certain databases, and the deficiency of cooperation with other agencies, especially the DMV, can negatively affect LPR utility. However, LPR technology is still highly valued by the departments RAND studied.

Although funding and resources are important concerns, apprehension over the privacy of and access to LPR data has caused several local and state lawmakers to closely examine LPR's role in law enforcement. This scrutiny and subsequent legislative actions has the ability to affect the expansion of LPR technology through the country and to affect how the systems are currently employed. The following chapter details privacy concerns and the legal history of the types of issues surrounding the use of LPR and LPR-like technologies.

CHAPTER FIVE

The Legal Aspect of LPR Privacy Concerns

License plate readers and the vast amounts of data they collect raise substantial privacy concerns. A police agency or private corporation conceivably could track the whereabouts of any particular automobile, providing information about a wide range of a person's private life, history, religion, or personal beliefs. As one court recently put it, "Disclosed in [location] data. . . will be trips the indisputably private nature of which takes little imagination to conjure: trips to the psychiatrist, the plastic surgeon, the abortion clinic, the AIDS treatment center, the strip club, the criminal defense attorney, the by-the-hour motel, the union meeting, the mosque, synagogue or church, the gay bar and on and on. . . ."[1] And as U.S. Supreme Court Justice Sonia Sotomayor has recently noted, the government "can store such records and efficiently mine them for information years into the future."[2]

Concerns about a surveillance state normally conjure fears of an intrusive *federal* government. Indeed, the checks and balances of the federal structure of our government were designed to minimize this risk. But the loss of locational privacy via LPR technology could also occur at a local or regional level. A handful of a system's cameras could quite effectively monitor nearly all travel in a rural area. An unscrupulous sheriff could monitor nearly all political activity by an opponent and quickly infer, based on the location of an associated license plate, who met with him or her. In this respect, the surveillance risks from this technology may be local and regional as well as national.

RAND's examination of the dimensions of this technology in the sections of this chapter reveal a fairly simple heuristic: The more closely that LPR systems can be framed as analogous to an officer keeping an eye out for a particular car or license plate associated with a specific crime, the more likely they will be to survive constitutional scrutiny and accord with our collective societal norms. But if the LPR systems seem to offer comprehensive surveillance and superhuman capacities, the more likely they will

---

1 *People v. Weaver*, 12 N.Y. 3d 533, 441-442, 309 N.E. 2d 1195,1199 (2009). Note: Court cases cited in this chapter are not included in the References.

2 *United States v. Jones*, 132 S.Ct. 945 (2012).

be found problematic. We consent to the possibility of being watched by a person (even a person with a tool), but an all-seeing machine is more problematic.[3]

Since the present study primarily addresses police use of license plate readers, we begin our discussion here with the Fourth Amendment to the United States Constitution, long considered the most critical protection for privacy from the government in the criminal context. As the U.S. Supreme Court explained: "The security of one's privacy against arbitrary intrusion by the police—which is at the core of the Fourth Amendment—is basic to a free society."[4] We then address state constitutional and statutory restrictions and other privacy concerns.

Fourth Amendment

The text of the Constitution's Fourth Amendment reads: "The right of the people to be secure in their persons, houses, papers, and effects, against unreasonable searches and seizures, shall not be violated, and no warrants shall issue, but upon probable cause, supported by Oath or affirmation, and particularly describing the place to be searched, and the persons or things to be seized." This amendment protects citizens against general searches and requires the government to establish probable cause, which specifically describes the things sought.

The Fourth Amendment's importance in the enforcement of criminal law has grown enormously due to two important 20th century developments. First, in *Weeks v. United States*,[5] the Supreme Court held that evidence seized in violation of the Fourth Amendment should be excluded from admission at trial, an application of the exclusionary rule—a penalty to deter law enforcement from violating the Constitution. Second, in *Mapp v. Ohio*,[6] the Supreme Court held that the Fourth Amendment applied to the states and that evidence seized illegally by state law enforcement also must be excluded from admission in state courts as a matter of federal constitutional law. As a result of these two cases, almost any evidence obtained as a result of a violation of the Fourth Amendment cannot be used to obtain a conviction. This makes the constraints imposed by the Fourth Amendment critically important in considerations of plate reader technology by police.

---

[3] Cf. Somini Sengupta, "Rise of Drones in U.S. Drives Efforts to Limit Police Use," *New York Times*, 15 February 2013.

[4] *Wolf v. Colorado*, 338 U.S. 25 (1949).

[5] *Weeks v. United States*, 232 U.S. 383 (1914).

[6] *Mapp v. Ohio*, 367 U.S. 643 (1961).

Contemporary Fourth Amendment Doctrine

While modern Fourth Amendment law has become a confusing mass of exceptions,[7] we will summarize basic doctrine here to provide a foundational level of knowledge. A critical question that has particular applicability to the issue of law-enforcement use of plate readers is whether their use constitutes a search at all.

In *Katz v. United States*,[8] the court held that a constitutionally relevant search occurs when the defendant has a reasonable expectation of privacy in the thing searched. In *Katz*, the FBI attached a listening device to the outside of a telephone booth from which the defendant booked bets. Based in part on this recording, the defendant was convicted of transmitting wagering information by telephone in violation of federal statute.

In its opinion, the court first explained that the Fourth Amendment was not synonymous with a generalized right to privacy:

> [T]he Fourth Amendment cannot be translated into a general constitutional "right to privacy." That Amendment protects individual privacy against certain kinds of government intrusion, but its protections go further, and often have nothing to do with privacy at all. Other provisions of the Constitution protect personal privacy from other forms of governmental invasion. But the protection of a person's general right to privacy—his right to be let alone by other people—is, like the protection of his property and of his very life, left largely to the law of the individual States.

The government argued that (1) because the phone booth was a public place, there could be no expectation of privacy in conversations therein, and (2) the Fourth Amendment did not apply because the surveillance technique did not involve any physical penetration of the telephone booth.

The court rejected both these arguments. From its perspective, the Constitution protects "people, not places," and the fact that the government did not trespass within the telephone booth was irrelevant. The court held that:

> [T]he Government's activities in electronically listening to and recording the petitioner's words violated the privacy upon which he justifiably relied while using the telephone booth and thus constituted a "search and seizure" within the meaning of the Fourth Amendment. The fact that the electronic device employed to achieve that end did not happen to penetrate the wall of the booth can have no constitutional significance.

---

[7] Akhil Reed Amar, "Fourth Amendment First Principles," 107 *Harv. L. Rev.* 757, 758 (1994). ("The result is a vast jumble of judicial pronouncements that is not merely complex and contradictory, but often perverse.")

[8] *Katz v. United States*, 389 U.S. 347 (1967).

Katz remains the law in defining whether law enforcement executed a search or seizure. So, under *Katz*, a key question about license plate readers is whether one has a reasonable expectation of privacy in one's locational privacy.

However, the court has generally held that when people reveal information to the public, they do not possess a privacy interest in it. So, for example, in *Katz* itself the court explained that "[w]hat a person knowingly exposes to the public, even in his own home or office, is not a subject of Fourth Amendment protection."[9] The court has applied this principle to find that people have no Fourth Amendment interest in trash they set out for pick-up,[10] or in the phone numbers dialed by a suspect, collected by a pen register at the telephone company.[11] Similarly, in *United States v. Miller*,[12] the Supreme Court rejected a Fourth Amendment challenge to the government's warrantless procurement of a defendant's bank account information. The defendant in that case argued that he had a reasonable expectation of privacy in his bank records because he only made them available to the banks for a limited purpose. The court rejected that argument and held that granting expectation of privacy entails the risk that the third party will reveal the bank account information to the government.[13]

Electronic Surveillance and Automobiles

The court considered the effect of electronic surveillance as applied to automobiles in 1983, in *United States v. Knotts*.[14] There the government installed a radio transmitter in a drum of chloroform to aid in tracking defendants, who were suspected in the manufacture of illegal drugs. Using the transmitter, police tracked them to a rural cabin. A search warrant was obtained to search the cabin and an illegal drug lab was discovered. The court unanimously held that no search or seizure occurred because a "person traveling in an automobile on public thoroughfares has no reasonable expectation of privacy in his movements from one place to another."[15] The defendant argued that "twenty-four hour surveillance of any citizen of this country will be possible, without judicial knowledge or supervision." The court noted that "if such dragnet type law-enforcement practices as respondent envisions should eventually occur, there will be time enough then to determine whether different constitutional principles may be applicable." Under the rationale of *Knotts*, license plate reader use is permissible because a person has no reasonable expectation of privacy on the open public roads.

---

[9] The *Katz* court implicitly found that speaking in an enclosed phone booth did not disclose this information to the public.

[10] *California v. Greenwood*, 486 U.S. 35 (1988).

[11] *Smith v. Maryland*, 442 U.S. 735 (1979).

[12] *United States v. Miller*, 425 U.S. 435 (1976).

[13] *Smith v. Maryland*.

[14] *United States v. Knotts*, 460 U.S. 276 (1983).

[15] *United States v. Knotts*.

More recently, however, the court appears to be grappling with the need to define a constitutional limit to electronic surveillance of travel. In 2012, justices decided *United States v. Jones*,[16] a case that brings up many of the same issues as license plate readers. In that case, the court addressed whether a Global Positioning System (GPS) tracking device attached to a car constituted a search. In *Jones*, the defendant was suspected of narcotics crimes and a small GPS device was attached to his Jeep Grand Cherokee. No warrant was obtained. The unit tracked the automobile's whereabouts 24 hours a day for four weeks or so.

The majority opinion, by Justice Antonin Scalia, found that "the Government's installation of a GPS device on a target's vehicle, and its use of that device to monitor the vehicle's movements, constitutes a 'search' under the Fourth Amendment." The court rejected the government argument (based on *Knotts*) that by driving on public roads, a person surrenders his privacy interest.

The court's holding, however, was fractured, so it is somewhat difficult to predict exactly how the justices would resolve the Fourth Amendment issues raised by LPRs. Justice Scalia's majority opinion turned on the government's installation of the GPS tracker constituting a trespass, an apparent return to the pre-*Katz* analysis of the defendant's property rights in the thing searched and whether the government committed a trespass.[17] Since the government action involved a trespass, it was unnecessary to address the more subjective, reasonable expectation-of-privacy test. Justice Alito wrote an opinion concurring in the judgment (joined by Ginsburg, Breyer, and Kagan) based on the theory that the government action violated the defendant's reasonable expectation of privacy under *Katz*. Per Alito's opinion, the trespass was legally irrelevant. He distinguished *Knotts* because of the comparatively short monitoring involved in that case. Justice Sotomayor published a concurring opinion in which she argued that "the Katz reasonable-expectation-of-privacy test augmented, but did not displace or diminish, the common-law trespassory test that preceded it."[18]

Despite the court's divergent opinions, there appear to be at least five votes (Alito and the three other justices who joined him, plus Sotomayor) for the position that, even absent a trespass, warrantless collection of location data over an extended period of time would constitute a search under the Fourth Amendment. So, for example, it seems likely that if a police force had a plate reader system (fixed or mobile) that tracked the location of an individual over an extended period, the court likely would find that a search occurred.

---

[16] *United States v. Jones*, 132 S.Ct. 945 (2012).

[17] Elsewhere, Justice Scalia, author of the majority opinion in *Jones*, has been very critical of *Katz* as having "no plausible foundation in the text of the Fourth Amendment," and "notoriously unhelpful" in identifying what constitutes a search. *Minnesota v. Carter*, 525 U.S. 83 (1998) (Scalia, J. concurring).

[18] *United States v. Jones*.

Some important questions, however, are unresolved. The discussion above indicates that the interpretation of whether a search has occurred depends on important details relevant to how LPR technology is used. The first is an issue of the duration of the tracking. Suppose that police simply tracked a vehicle with the technology for a limited period—perhaps three days. It is unclear whether this would be analogous to the limited tracking the court found acceptable in *Knotts* or to the more extended monitoring in *Jones*. When the system is mobile (on vehicles) or fixed (on buildings or bridges), it presumably would take much more than a few days to gather enough location data about an individual vehicle to constitute "tracking." If, however, a unit is mounted on a portable trailer or parked vehicle deliberately positioned to target a specific vehicle or set of vehicles, an argument of tracking might be more valid in a shorter time.

Similarly, the question of the pervasiveness of the tracking is also unresolved. When deployed on mobile vehicles or fixed infrastructure, license plate readers are not focused on an individual car but rather on every vehicle within range. In the absence of a comprehensive system, it is probable that hits for a particular car are likely to be haphazard and limited to whenever the target happened to be picked up rather than comprehensive. This is quite unlike the monitoring in *Jones*, which was limited to one car and was comprehensive. However, when a portable plate reader unit is deliberately positioning to target specific vehicles, that use becomes more similar to what might constitute a search.

Given the rapid adoption by law enforcement of these systems, as chronicled by this report, it is almost certain that defendants will bring Fourth Amendment litigation challenging the use of this technology. Such litigation likely will reach the Supreme Court eventually and should clarify the Fourth Amendment's restrictions (if any) on the use of license plate readers.

State Constitutional Restrictions

Some state supreme courts have also interpreted their state constitutions to restrict technology that can infringe upon locational privacy. The Oregon Supreme Court held that the warrantless use of a beeper that tracked the defendant's automobile deprived the criminal defendant of his privacy rights.[19] Similarly, the Supreme Court of the State of Washington also has expressed concern about this technology:

> [U]se of GPS tracking devices is a particularly intrusive method of surveillance, making it possible to acquire an enormous amount of personal information about the citizen under circumstances where the individual is unaware that every single

---

[19] *State v. Campbell*, 759 P.2d 1040, 1048-49 (Or. 1988) (Absent a warrant requirement, "no movement, no location and no conversation in a 'public place' would in any measure be secure from the prying of the government. There would in addition be no ready means for individual to ascertain when they were being scrutinized and when they were not. That is nothing short of a staggering limitation upon personal freedom.")

> vehicle trip taken and the duration of every single stop may be recorded by the government.
>
> We conclude that the citizens of this State have a right to be free from the type of governmental intrusion that occurs when a GPS device is attached to a citizen's vehicle, regardless of reduced privacy expectations due to advances in technology.[20]

Subsequent litigation in the state courts will likely clarify the scope of these state protections.

State Regulation of License Plate Reader Use

Specific Prohibitions

The state of New Hampshire has enacted a law that specifically prohibits law-enforcement officers from using "automated number plate scanning devices" outside of certain narrowly specified circumstances.[21] A different law generally prohibits "highway surveillance," which is defined as

> the act of determining the ownership of a motor vehicle or the identity of a motor vehicle's occupants on the public ways of the state or its political subdivisions through the use of a camera or other imaging device or any other device, including but not limited to a transponder, cellular telephone, global positioning satellite, or radio frequency identification device, that by itself or in conjunction with other devices or information can be used to determine the ownership of a motor vehicle or the identity of a motor vehicle' s occupants.[22]

The statute exempts surveillance if it is specifically authorized by statute; undertaken on a case-by-case basis in the investigation of a particular violation, misdemeanor, or felony; or other relatively narrow exceptions.[23]

The state of Maine also has specifically prohibited the general use of "automated license plate recognition systems."[24] Its ban extends to private parties using such systems but does not apply to (1) the Maine Department of Transportation for the protection of public safety and transportation infrastructure, (2) the Department of Public Safety, Bureau of State Police for the purposes of commercial motor vehicle screening

---

[20] *State v. Jackson*, 76 P.3d 217, 223-24 (Wash. 2003). The court upheld the use of this surveillance because a warrant was obtained.

[21] N.H. Rev. Stat. Ann. § 261:75-b (2013).

[22] N.H. Rev. Stat. Ann. § 236:130.

[23] N.H. Rev. Stat. Ann. § 236:130.

[24] 29-A Maine Revised Statutes, Annotated (M.R.S.A), § 2117-A.

and inspection, and (3) any state, county, or municipal law-enforcement agency when providing public safety, conducting criminal investigation, and ensuring compliance with local, state, and federal laws. However, law-enforcement use of license plate readers has to be "based on specific and articulable facts of a concern for safety, wrongdoing, or a criminal investigation or pursuant to a civil order or records from the National Crime Information Center database or an official published law-enforcement bulletin."[25] The act also requires that license plate reader data be kept confidential and discarded after 21 days if it is not considered "intelligence and investigative information."

More-General State Restrictions on Data Collection

Other states have more-general restrictions on government collection and retention of data. While a comprehensive review of all state restrictions on data collection is beyond the scope of this report, we will discuss Virginia as one example, because the Virginia State Police recently sought the state attorney general's position on the legality of license plate readers under Virginia's Government Data Collection and Dissemination Practices Act ("Data Act"), which regulates certain kinds of data collection by the Commonwealth of Virginia and its entities. It is therefore an interesting example of a state legal limit on this technology's use. We did not survey states to determine whether the Virginia case is representative, but we use it as an illustration.

In 2012, Colonel W. S. Flaherty, the superintendent of the Virginia Department of State Police, sought a formal, legal opinion from the Virginia State Attorney General as to whether the use of license plate readers violated this law. While the legislation was deemed target-specific, the law also could be seen to apply to plate data, among other types. In a February 13, 2013, letter,[26] Attorney General Kenneth T. Cuccinelli opined that the Data Act restricted license plate readers' use by the government. The Opinion Letter first noted that data collected using such systems fell under the Data Act's definition of personal information and was therefore restricted.

The opinion then considered whether use of such LPR systems met any exceptions in the Data Act for law enforcement. The act exempts data used by police departments "that deal with investigations and intelligence gathering related to criminal activity" from its general restrictions, so the question is whether the use of license plate readers falls under that exemption.

The attorney general distinguishes the "active" use of license plate readers to look for a particular plate involved in criminal activity from the "passive" use of such systems to collect data more generally. He finds that "active" use of license plate readers does not run afoul of the Data Act:

---

[25] 29-A M.R.S.A. § 2117-A.

[26] Kenneth T. Cuccinelli, Virginia attorney general, letter to Col. W.S. Flaherty, superintendent, Virginia Department of State Police, 13 February 2013.

> Clearly, data collected by an LPR in the active manner and maintained by such law enforcement entities relates directly to the immediate public safety threat of criminal activity. Thus, such data is exempted from the application of the Data Act by its specific terms.[27]

In contrast, he finds that the passive use of license plate readers runs afoul of the Data Act and cites section 2.2-3800(C) of the Code of Virginia, which indicates that "[i]nformation shall not be collected unless the need for it has been clearly established in advance." The Opinion Letter states that the passive use of license plate readers results in data for which the need has not been established in advance. "Its future value to any investigation of criminal activity is wholly speculative." As a result, he finds that "the collection of LPR data in the passive manner does not comport with the Data Act's strictures and prohibitions, and may not lawfully be done."[28]

The dichotomy between "active" use and "passive" use seems problematic. Data collection by plate reader units is not selective and so the distinction between active and passive use applies not to data collection but to its storage. Whether such data "deal with investigations and intelligence gathering related to criminal activity" can only be determined after the data are already collected. This creates a catch-22 situation since the law-enforcement agency can only determine if the information relates to an "immediate public safety threat of criminal activity" and is therefore eligible for collection after the data have already been collected.

Substantively, the distinction also may be hard to make. Suppose, for example, that data are stored for one hour while a complex analytic program scours the information collected by multiple plate readers for patterns consistent with criminal activity. Would such use be deemed passive or active? Or suppose that the system is set up to provide alerts about not just to cars tied directly to crimes but also to license plates that may (or may not) be indirectly associated with criminal activity—say, that are registered to the household of a once-convicted felon. Would such use be considered active or passive? There are some indicia of particularized suspicion, but they are relatively weak. In short, the active-passive use dichotomy on which the attorney general's opinion is based seems potentially problematic. It will be interesting to see if courts adopt it as a way of distinguishing permissible use of this technology from that which is not.

Liability Risk in Use

The ACLU, as noted in Chapter Two, has been active in seeking information on the extent of LPR use. Generally, this has involved filing federal Freedom of Informa-

---

27 Cuccinelli, p. 3.

28 Cuccinelli, p. 4.

tion Act requests or their state law analogs and litigating any denials. These suits have sought to gather information on the amount and scope of data collected rather than any direct effort to halt the use of the technology or to collect damages from an agency or department.

Because even the extensive use of license plate readers is legal in most states, it is unlikely that any public (or even private) users of such systems would face civil liability. Even if a litigant could identify a legal theory to support a lawsuit, the doctrines of sovereign and qualified immunity would pose additional obstacles to a successful suit, at least under current law in most states.

Nevertheless, because of the privacy implications of this information, it would be wise for most departments to establish clear policies regarding data retention, access, and compliance with these policies. Doing so would help address some of the privacy concerns raised by this technology and minimize the likelihood of a lawsuit.

Constitutionally Relevant Dimensions

The use of license plate readers raises substantial privacy concerns. The Supreme Court has strongly suggested that it would consider pervasive monitoring with LPR systems to be a search subject to the Fourth Amendment, but it left many unanswered questions about exactly when law-enforcement use of plate readers likely violates the Fourth Amendment.

One way to think about this is to imagine a continuum of police use of license plate readers, from those acceptable under the Fourth Amendment to those almost certainly not. In general, the more closely the technology's use is tied to a specific criminal investigation of a particular suspect, the more likely it will survive constitutional scrutiny; the more open-ended and pervasive law-enforcement use of it is, the less likely its use will survive Fourth Amendment scrutiny.

We can identify several dimensions along which law-enforcement use of the technology might vary and which may affect the likely constitutionality of the monitoring.

Geographic Comprehensiveness

If a comprehensive network of license plate readers is set up so that virtually any travel is subject to monitoring, as opposed to a more limited system and monitoring, it is more likely that courts will find searches using such a network in violation of the Fourth Amendment and in need of a warrant.

Use Tied to a Specific Criminal Investigation

Law-enforcement searches for a particular license plate closely related to a specific crime are more likely to survive constitutional scrutiny than more open-ended hunts for criminal activity patterns or scours for plates more distantly connected with a suspect. Suppose, for example, that the police search data pulled from a comprehensive LPR network for patterns of movement by friends and family members of a possible suspect and this leads to incriminating evidence. This is less likely to survive constitutional scrutiny.

Real-Time Use Versus Data Mining

In general, the more contemporaneously the license plate reader information is used, the more closely it resembles the constitutionally long-accepted practice of an officer keeping an eye out for a particular vehicle and the more likely its use will be upheld. This corresponds to the "active" use terminology used by the Virginia attorney general as discussed above. In contrast, the "passive" collection of larger amounts of license plate reader data and subsequent analysis of the information is likely to face more constitutional scrutiny. Since several departments interviewed for this study specifically mentioned "data mining," a passive use of plate reader data and a capability they valued, this issue may be tested in the courts in the near future.

Government Owned and Operated Versus Third-Party Owned and Operated

License plate reader data are more likely to survive constitutional scrutiny if collected by private third parties and then obtained by the government via subpoena or request. Under long-standing Fourth Amendment doctrine, information voluntarily disclosed to a third party loses its constitutional protections.[29] By driving on the public road on which a private third party operating a plate system can legally collect data, a criminal defendant voluntarily shares his locational information with the third party and therefore possesses no constitutional right in it.

Conclusion

The Fourth Amendment, at least as currently interpreted, is an imperfect means to protect privacy interests, in large part because it is usually "enforced" by criminal defendants seeking suppression of incriminating evidence. Although accused defendants have considerable interests in vigorously advocating for their privacy rights, they represent a relatively atypical fraction of the U.S. population. Moreover, the fact patterns that arise in criminal litigation do not span the gamut of situations in which we might be concerned about privacy. Finally, the Fourth Amendment only protects citizens against the government use of information and has no effect on potentially pervasive third-party collection of information, including private operators of license plate readers.[30]

Broader statutory efforts to protect privacy also will affect license plate reader use, as has occurred in Virginia. As the use of this technology receives more publicity, it is possible that more states will pass laws governing its use and the storage of the

---

[29] The two cases that established this doctrine are *United States v. Miller* and *Smith v. Maryland*.

[30] Angwin and Valentino-Devries (private companies have photographs of a large majority of registered vehicles).

large amounts of data collected. There also is growing awareness of the general issue of locational data privacy and the ways that an individual's location can be tracked by cell phone tower geolocation and the fact that many smart phone and mobile device applications make use of GPS data on phones. Similarly, autonomous vehicle technology may use locational data. These growing uses and awareness of locational data issues could lead to more general policymaker efforts to restrict or govern locational privacy issues that are likely to affect the use of license plate readers.[31]

Alternatively, it is possible that we will become so accustomed to the loss of locational privacy that it will not seem worth protecting. This may occur, some have hypothesized, because younger generations have become so habituated to sharing locational and other private information via mobile phone applications. This kind of changing norm is incorporated into Fourth Amendment analysis. If one no longer has a "reasonable expectation" of locational privacy, then the Fourth Amendment (at least as interpreted under *Katz*) will no longer protect it. Hard choices will have to be made collectively about the tradeoffs between the benefits from law-enforcement use of this technology and the loss of locational privacy that may result.

---

[31] In France, for example, the European Data Union Protection Directive was interpreted to protect the privacy of a license plate number. Déliberation No. 96-069 du 10 septembre 1996 relative à la demande d'avis portant création à titre expérimental d'un traitement aautomatisé d'informations nominatives ayaant pour finalité principale la lecture automatique des plaques d'immatriculation des véhicules en movement par la société des autoroutes Paris-Rhin-Rhône (SAPR).

CHAPTER SIX

Common Themes and Challenges in LPR Use

An important characteristic of RAND's research was the analysis of case study data from multiple perspectives. We present our findings as themes that surfaced from comments in the case studies, with data access and information retention policies having a highlighted role. We discuss technology-driven themes, operationally driven themes, privacy, and the benefits and challenges of the technology.

It is important to remember that the key finding regarding the information RAND collected through the case studies is that LPR is useful for any type of case, depending on the support the system has inside the department. And, as stated in Chapter One, the department must have access to relevant databases and data retention policies sufficient to fully exploit the data. The subsequent themes presented in this chapter present the specific evidence that forms the basis for this key finding.

Technology-Driven Themes

LPR Utility Depends on Data, Human Resources

Data

As with any data management system, LPR's utility depends strongly on the type and quality of the data with which it works. "Hotlists" are built with data compiled from multiple sources, including the FBI's National Crime Information Center, state law-enforcement agencies, state motor vehicle departments, and local law-enforcement agencies. Access to these data sources varies from agency to agency.

The NCIC prepares a special file for plate reader systems with vehicle information from a number of its standard files (e.g., stolen vehicles, wanted persons, missing persons, and several others). This file is compiled, refreshed twice daily, and made available to most state and a small number of local law-enforcement agencies.[1] State agencies often add analogous vehicle data associated with state-level criminal information data-

[1] "License Plate Reader Technology Enhances the Identification, Recovery of Stolen Vehicles," CJIS Link, Federal Bureau of Investigation, United States Department of Justice, Washington, DC, 2011.

bases and make this combined file available to local agencies. All the agencies we met with had access to such data files from their states.

Access to state motor vehicle department data was less uniform. Some agencies received twice-daily updates from state motor vehicle departments of vehicle data associated with expired registrations, suspended driver licenses, and other violations. However, agencies we met with in some states lacked access to this information, either because data-sharing agreements had not been worked out or because questions about who would pay the cost to compile and update data had yet to be resolved.

Because the volume of expired registrations and suspended licenses plates is so great, whether agencies have access to state motor vehicle department data had a strong influence on their activity with plate reader systems. These data are the source of the vast majority of system alerts, say agencies with access to the information.[2] In some cases, alerts were so frequent that officers had to ignore them or to adjust the system settings to neglect matches to expired registrations and suspended licenses, which one official said in an interview set off an alert every few seconds, making it impossible to respond to them all. One agency deliberately did not use these data because it felt it lacked the resources to respond to the alerts they would prompt; with information from the NCIC, state law enforcement, and its own locally generated data, it had sufficient material and alerts to respond to already.

Finally, agencies also add data on their own. These generally are either urgent additions from national or state LPR systems received between the regular twice-daily updates or data provided by their own agency associated with local criminal activity.

NCIC and state data were downloaded to local agency computer LPR systems via standard file transfer protocol (FTP) or secure email. From there, all but one agency transferred data to and from vehicles though mobile wireless networks. One small agency with few plate reader units said it manually transferred data to and from vehicles with a USB flash drive. Once data were in the plate reader units, the system automatically could combine separate information streams into a single hotlist against which scanned plates were compared.

The practicality of manual data transfer is clearly limited to agencies with only a small number of LPR units. In addition, manual data transfer hampers the ability to add urgent license plate data arising between normal data transfer times. Such additions must be relayed over the radio and manually entered by vehicle drivers.

Human Resources

As with any new technology, integrating LPR systems requires commitment and patience. Several agency representatives discussed technical challenges in implementing LPR systems; these primarily were associated with understanding and adjusting settings for server software. Administrators, they said, must be willing to troubleshoot

[2] An alert is a match between a license plate read by the LPR camera and a license plate in the hotlist.

and tailor systems to an agency's needs. They may need to address hardware issues, dig into the software, and develop vendor relationships. They also must be willing and able to train officers to use the system. Agencies generally appeared capable of identifying staff with the necessary skill and perseverance to set up LPR systems, run them, and train others. Who took on these roles varied by agency. Agencies also varied in whom they assigned to keep, analyze, and prepare operational statistics on plate reader systems; this requires yet more time and staffing commitment.

Fostering the necessary dedication and enthusiasm by administrative staff required, in turn, commitment and resources from agency leadership. Leadership commitment facilitated the integration of LPR use throughout the agency. Representatives noted that officers differed in their enthusiasm about using LPR. A strong signal from the chief about the importance of LPR was observed to help encourage ambivalent officers.

All the agencies we met with had acquired their systems with Homeland Security grants, usually as part of a large regional award. In some cases, an agency had little to do with the decision and simply was provided funding for a system. This did not seem to affect the agencies we met with, but it might create problems if a department is not actively seeking a system but gets one and no one takes ownership of it or champions its use and integration.

Interoperability Among Jurisdictions Requires Several Elements

Because vehicles travel freely among different jurisdictions, it is important that LPR systems can share data among law-enforcement agencies. While national and state hotlist data are available to many, agencies near state (or national) borders likely will encounter plates from areas nearby, and it would be mutually beneficial for those in a region to share their locally generated hotlists as well as plate reads. For investigative purposes, quick and convenient access to sightings of particular license plates by other agencies' systems was particularly helpful.

Because LPR systems are, for the most part, owned and operated by individual agencies, sharing data requires added effort beyond normal operation. Based on our discussions with agency representatives, we identified three elements required for interagency data sharing.

The first, simply, was availability of data. Agencies in some states did not yet have access to state motor vehicle department data. These data could not be shared with agencies in other states unless a means to share data to agencies within the state was worked out.

The second element was the technical interoperability of LPR systems among agencies wishing to share data. As with any new technology, different suppliers use varying data formats and transfer protocols that make sharing difficult. As technologies mature, markets typically converge toward standardized formats, easing interoperability. RAND's cases involve seven agencies; accordingly, we offer a limited view on system interoperability. Agency representatives confirmed that it was only possible to share data with agencies using the same brand of LPR system. They also indicated that, while there are a number of suppliers, the market is largely dominated by two large suppliers and most agencies within a given regional area generally use a single supplier. Some agencies successfully shared data with neighbors. This suggests that technical interoperability may not hamper cooperation, especially due to system vendor dominance. A more comprehensive examination could confirm this.

A final element of sharing LPR data was developing administrative agreements between agencies. Several departments we met with identified these agreements as the primary impediment to information-sharing. It appears to be challenging to find resources to put together the needed technical, legal, and consensus-building expertise to create these agreements. Some agencies foundered on drafting an entirely new agreement just for this technology; they preferred to develop a general data-sharing agreement that applied to all such types of law-enforcement information, with the option to add specific data types and technologies as the need arose.

Despite these challenges, some agencies we met with had put in place agreements and were sharing data, including an instance where more than 80 agencies signed on. While that arrangement was initiated and managed mostly by a single large agency, each participant had direct access to the shared database.

Data Storage and Retention: Technology Drives

Technical constraints drive how long agencies can store LPR data, according to representatives of most of the agencies with whom we met. Several said that privacy concerns weighed heavily as they first considered how long to keep their license plate data. No duration for data storage was universally accepted, especially to appease privacy concerns; the agencies we interviewed said they intended to store data for one to two years, based largely on peers' practices. Agencies soon learned that storage space was the bigger constraint on data retention. Some agencies found they lacked the resources to keep two years' or even one year's worth and they cut their storage. Some agencies saved records that do not match a hotlist (sometimes called "reads") for six months and retained those matching a hotlist ("alerts" or "hits") for 12 to 24 months. One agency lacking a central computer server stores its plate data for only 90 days because the information is kept in system units in patrol vehicles. In all cases, data potentially tied to an active investigation are kept as long as necessary.

An agency with particular interest in cold cases developed a policy for unlimited storage of system data for possible use in future investigations (including protocols for data security and access limitations). That policy had yet to be implemented because necessary technical capacity was unavailable.

Operationally Driven Themes

LPR Funding Is Largely External

With most police departments confronting shrinking budgets and cuts, funding new technologies can be problematic, forcing agencies to conduct cost-benefit analyses when acquiring new LPR systems and equipment. In the case of LPR technology for agencies in our sample, they did not need to come up with big sums from their own budgets because most of the funding for this technology came externally. Agencies have received significant state and federal grants for counterterrorism and security initiatives, and these have let them upgrade their technology, including by buying LPR systems.

The range of federal and state funding varied by department. Several benefited from Department of Homeland Security Urban Area Security Initiative grants. This program addresses "the unique planning, organization, equipment, training, and exercise needs of high-threat, high-density urban areas, and assists them in building an enhanced and sustainable capacity to prevent, protect against, mitigate, respond to, and recover from acts of terrorism."[3] North Town received funding from its state homeland security agency to upgrade security at multiple industrial sites throughout the region. It also received funding from state agencies, which approached multiple jurisdictions with sums for a range of technological and tactical improvements, including LPR systems. South City received state funding under initiatives to improve border security. The departments in the cluster category in this report received funding from a local governmental cooperative.

In some cases, funding was designated to purchase LPR systems; in other cases, agencies had flexibility in how to spend their grants. In general, departments tapped grants to buy their first plate reader cameras and other system infrastructure and support. Many systems were supported entirely by grants. But in most cases, those grants have expired, and many agencies worry that they now must find new sources of money to support this technology. Several said that costs and lack of equipment warranties limited their ability to maintain system cameras; others said they could not expand their LPR systems because of budget constraints. So while experiences with the technology have been positive, it seems unlikely that there will be more equipment buys to expand networks.

Several agencies discussed spreading costs for systems' maintenance and expansion among departmental partners and municipal entities. Big City plans to charge members in its regional law-enforcement cooperative a fee to be applied to the cost of data storage. South City may partner with the local municipal courts to provide overtime to officers to use its plate system to identify traffic scofflaws who owe court fines, with sums collected split between them. The police would use their share to maintain

[3] Federal Emergency Management Agency, FY 2012 Homeland Security Grant Program web page.

and expand their plate system. With tight budgets the likely norm for a time, spreading costs and developing strategies to raise revenues to maintain LPR systems are likely efforts for many departments.

Major Case Uses for LPR: Often Department-Specific

The types of investigations for which each department employs LPR systems appear to be distinctive to the operational environment and resources available. Auto theft continues to be a major use for the systems nationwide. Cluster City and South City stated this is so for them. South City's system helps not only to identify stolen vehicles but also to uncover evidence tied to drug-related crime. Cluster City attributes 90 percent of its stolen vehicles recoveries to its system. Other departments, however, reported that few stolen vehicles were recovered using LPR. It appears the environment in which systems operate affects their utility in these cases. So LPR systems may often be useful for auto theft, but they do not guarantee increased rates of stolen vehicle recovery.

Evidence from our interviews suggests that access to state DMV data influences the technology's utility. Both DMV-Enabled Cluster County and North Town said that a major driver in their LPR systems' use was identifying DMV-related violations, with suspended driver's licenses and expired tags making up a large portion of their application. Cluster City, Cluster County, Small Town, and South City did not use their LPR systems for DMV violations because they did not have or did not want access to that type of data; several find non-use for this type of case hindered their LPR systems' overall utility. Only one department, Big City, did not generally pursue those alerts even with access to DMV data, stating that enough more-serious alerts existed to occupy the time of system users. As with auto theft, the resources available to a given agency appear to drive whether it uses the technology for this case use.

The technology's efficacy in ongoing investigations was another major theme of operational use. Most departments said they could access their LPR systems' databases to assist detectives in a range of criminal investigations. This data retrieval could provide information on vehicles at a crime scene, patterns of suspects, or corroboration of alibis. But access to stored data varied: Some agencies granted it to all investigators; in others, only a limited number of officers have access and investigators must make specific requests. Data from other agencies were normally only available to officers who called system administrators, who then printed the information out and passed it to investigators; staffing affected the time between such requests and when they were fulfilled. Big City and South City participate in regional fusion centers staffed by analysts with access to the database, and this expedited the passage of information to the field. Several agencies actively were developing means for direct access to their neighbors' system servers, though this access has been limited until now.

LPR data generally have not been mined and analyzed for predictive policing or intelligence-led operations in our case study agencies. In theory, storing system data for long periods could let analysts develop clues about where criminal activity occurs.

Cluster City is working with its public works department to create maps with its plate system data to determine auto theft hot spots. There were few other reported instances of the data helping to predict crime patterns. Several agencies said it would be desirable to use the data this way, but limits on personnel and other resources prevented this. As was observed with incorporating DMV-related violations, taking advantage of system data to assist ongoing investigations or predictive policing was constrained by the resources of the individual departments and their data-retention policies.

System Structure and Policies Vary

The type of cameras, where they are placed, and the structure of the supporting database affects LPR system use, as do agencies' policies and procedures for investigating their alerts and storing system data as evidence. The cameras can be fixed, vehicle-mounted, or portable. In all the agencies we met with, they were attached to police cruisers, meaning they could go to any area police target in the jurisdiction. Fixed LPR systems typically go on existing structures and require independent power sources. Several agencies ran portable trailer units requiring similar infrastructure support; these also could be moved as circumstances dictated. South City has suitcase-sized portable units with small cameras. The units can easily be relocated, and they attach to vehicles with magnets. Each system has advantages and disadvantages, which each agency takes into account.

Mobile LPR systems are cheaper than fixed units, and they do not require existing infrastructure for support. But this use can have its drawbacks. Units may go out of service and their valuable field service time may be lost if they are attached to vehicles requiring maintenance. LPR systems also suffer when officers less technologically savvy are assigned to camera-equipped vehicles. These officers, anecdotal information indicates, were less likely to use the technology than those comfortable in adopting new technology.

Fixed LPR systems have an advantage in that they are operational at all times, meaning they have more reads and alerts than mobile units. But fixed cameras must be installed on structures by staff and require further support for power and data transmission. They also require dedicated staff to monitor and confirm alerts and initiate responses (i.e., to dispatch patrol cars, notify investigators, etc.). Several agencies encountered challenges in securing sites for fixed plate-reader cameras, co-locating them with existing infrastructure and working with local electric companies or private property owners to power their system and to fill fixed units' unique maintenance requirements (e.g., some could be reached only with bucket lift trucks).

Each agency decides the types of cameras it deploys; mobile cameras units are most prevalent because they are cheaper and easier to maintain. Several departments noted privacy concerns over fixed cameras because they read every plate, 24 hours a day at the same location, and thus might more easily track individuals. Officials from South City rejected fixed cameras because of privacy concerns. But Cluster City operated 38 of them in a jurisdiction with extensive, non-LPR surveillance assets for law enforcement and counterterrorism. Issues with connecting fixed cameras with existing infrastructure also contributed to less use of them than mobile cameras.

Policies on LPR use appeared more uniform across agencies, with only a few variations. Those with many commonalities included procedures for investigating an alert, assigning vehicles, and storing system information as evidence. All the departments relied on officer discretion to determine which alerts to pursue. LPR technology can increase officers' workload substantially when operational. One department noted that an officer received 11 alerts in one minute while passing a row of cars; obviously, it was impossible for the officer to pursue every alert. More-serious crime receives priority, while officers use their discretion about many DMV-related offenses. All but one department requires verification of an alert before officers take action. These departments do not consider system alerts as probable cause and require confirmation in a radio call to dispatch or confirmation on an officer's mobile terminal before a stop may occur. A key reason to confirm alerts is that hotlists in vehicles may be hours old and some plates no longer may be on the current list (e.g., if a stolen car is now recovered and is being driven by the owner).

There were commonalities among departments in the assignment protocols of vehicles with plate-reader units. In some agencies, these vehicles were older and less sought after by some officers. Camera-equipped vehicles also assigned as "take home" vehicles were also subject to increased downtime; these units thus became available only one shift per day. Many departments, therefore, assign camera-bearing cars to the pool of cars used each shift. Big City found that some older, less technology-savvy officers were not as enthusiastic about learning a new technology and did not take full advantage of the cameras. Therefore, camera cars often were assigned to younger, more technologically enthusiastic officers, who used the equipment more frequently.

Special procedures for using system information as evidence did not appear widespread. Only two departments noted instances of such information in the evidentiary process. South City said that if such information were to be evidence in a criminal prosecution, the specific reads were removed from the database and stored electronically under departmental rules on evidence handling. That information could then be stored indefinitely. Another agency said it had no special evidence-handling requirements for this information.

Privacy Concerns

Our research revealed significant privacy concerns regarding the use and storage of LPR data. These concerns concentrated on whether or not the data retained in LPR records was considered PII, how long the data were retained, and who had access to the information. The interviews found that department policies and practices related to LPR use were, in some ways, shaped by concerns about protecting people's privacy. However, while there is a general concern about privacy, there is great uncertainty about privacy expectations and the acceptable limits of LPR use. Consequently, departments noted that they had little information or guidance about how to incorporate privacy protections and were essentially improvising on their own.

The concept of PII was not widespread during the interviews. The general attitudes across departments appeared to be that LPR records only pertained to license plates and any additional information on the driver or owner of the vehicle required additional investigation. Additionally, there did not appear to be much concern over Fourth Amendment issues, likely indicating the departments felt that information gathered from license plates in public was an acceptable use of the technology. As the results from the George Mason survey indicated would be the case, many of the departments did not research or consult with departmental legal counsels on privacy implications of LPR systems. Since privacy issues were known at implementation, this suggests that departmental knowledge or consideration of LPR/privacy issues is not mature enough to offer effective guidance.

There was also uncertainty over determining how long LPR data are stored. Law-enforcement agencies preferred to keep LPR data as long as possible because data may turn out to be helpful in future investigations. Agencies are aware, however, that extended data storage raises privacy concerns, and so they have chosen to limit the amount of time that data not associated with an active investigation are kept. When asked how they decided how long to store it, agencies tended to examine what other agencies were doing, but generally without citing any clear link to experience or principles related to privacy per se. For the time being, data storage was often limited more by technical and financial constraints than privacy concerns (see above), although such constraints have been rapidly eroding as data storage becomes more affordable. Privacy considerations are therefore expected to become all the more important in determining how long LPR data are stored. The advent of cloud computing and remote data storage potentially further complicates this question by distancing agencies from their data and decreasing their control over this information.

Some departments have implemented strict controls on accessing the server data and have limited the number of people granted such access; in other departments, access was made available to essentially anyone willing to learn how to navigate the server data system. At least in some cases, server access was also limited by the resources available to register and train users. Departments strongly emphasized during training that LPR data should only be used for law-enforcement purposes. Another privacy-related area where departments differed from each other is the extent to which they publicized the use of LPR. One department, in an effort to assuage concerns about privacy and transparency, launched a "media blitz" to show off the new LPR system and answer questions from the media and public. In contrast, another department chose to minimize attention to the fact that it uses LPR in order to prevent "educating the criminals" about how to defeat LPR system capabilities.

Benefits and Challenges

Benefits

LPR systems have distinct benefits that make them attractive to police, including helping to more efficiently allocate resources, increasing opportunities for high-profile arrests, and boosting the capacity to identify those vehicles possibly involved in criminal activity, even with partial plate information. These benefits make LPR systems a force-multiplier for police.

LPR systems allow more efficient resource allocation, permitting patrol officers to concentrate on other duties while the system scans for vehicles of interest. LPR systems can scan hundreds of plates per hour, far more than an officer could input manually. The cameras can go into targeted areas, especially hot spots. In one agency's operation, officers put a system camera outside an apartment complex known for its involvement in auto theft. Several suspects were arrested because the cameras alerted officers nearby that stolen vehicles were going into the complex; police swooped in and caught suspects still in the vehicles. Other departments fix units in potential high-crime areas in lieu of patrols. The cameras note vehicles of interest in the area and discern patterns, allowing agencies to decide if they need to target resources there.

Several departments said the number of jailable offenses derived from traffic stops increases when plate-reader technology is in use. Few agencies have performed quantitative analysis to prove this, but one retrieved data attributing to its system a tenfold increase in the number of jailable offenses from traffic stops. Because courts issue warrants for people and not necessarily for vehicles, not all warrants are identified by plate system alerts. Officers sometimes will run background checks on drivers in a traffic stop and discover more outstanding warrants. Apprehensions for outstanding warrants associated with traffic offenses, such as hit-and-run and driving while intoxicated (DWI) offenses, increased because LPR systems identify vastly more vehicles of interest than officers can manually.

Photographs provided by LPR systems also assist in identifying vehicles. If an alert goes out on a stolen plate, the receiving officer can call and compare a photo to a vehicle description to determine the status of an offense. Photos also can provide information on vehicle drivers and passengers. If witnesses provide only partial plate information, this still can be run in LPR systems' database; a list of possible matches, combined with stored photos, can help police narrow possible suspect vehicles. This helps officers cut investigation time and allows for a more efficient use of resources.

Challenges

Many challenges for LPR systems have been addressed throughout this report, including cost, privacy concerns, and determining the right officers to use the equipment and the lack of resources to fully take advantage of data. These all are important when agencies consider whether to buy and deploy LPR systems. Besides these and other matters covered elsewhere, it is worth mentioning issues concerning standardization of records in the database and accuracy of the system.

Several agencies saw record standardization as an impediment to LPR systems' interoperability. Differences in type, format, and quantity of information on such records hindered sharing of information. One LPR systems administrator called for records to be standardized nationwide, emphasizing that this could be done with relative ease and to great benefit to LPR users in streamlining information-sharing.

Although this report does not delve into the technical competency of LPR systems nor of individual vendors, agencies in interviews highlighted quirks worth mentioning. LPR cameras and their processing software cannot distinguish between license plates from different states. This means that cameras might match a plate photo to a hotlist alert—but the plate may belong to a vehicle from the wrong state. The cameras also can false-read structures as license plates, as one department found when its system kept seeing wrought-iron fences around some homes as "111-1111" plates. Most agencies require alert verification via mobile terminals or confirming calls to dispatch before officers act.

CHAPTER SEVEN

Agency Lessons Learned

Here are a number of agencies' self-identified "lessons learned." They are classified as either

- actions/knowledge agencies believed would have been helpful to undertake/understand when they first adopted this technology, and/or
- concepts they believed would be helpful for other agencies to consider as they decide to adopt and deploy the technology.

We summarize these lessons and offer commentary on a number of them, describing benefits or potential areas of concern.

Estimate and Secure the Proper Amount of Funding Required

Explore Grant Options

Federal grants paid for the vast majority of LPR systems for the agencies RAND interviewed for this study; most of the grants came from Homeland Security for anti-terrorism applications. As such, local and state agencies may benefit from investigating the availability of such funding before planning system purchases; several jurisdictions within a region, facing common problems, could collaborate to apply jointly for a grant and, if successful, make a single, bulk purchase, lowering costs and overhead to each agency.

Be Prepared to Pay for Operations and Maintenance

Many agencies urged others to plan, ahead of purchase, for the funding needed to install, operate, maintain, and expand LPR systems. Some agencies, beneficiaries in some cases of group applications led by others, went ahead and bought systems without fully considering how they would pay to run and maintain them for their entire life-cycle. LPR systems are technology-based equipment that require updates and repairs. LPRs are of no use to the department if the department cannot fund repairs.

With the end of initial grant funding looming for several agencies, they scrambled to find ways to keep paying for continuing and variable costs, including data storage and system repairs. Agencies offered ideas for different ways to avoid this situation, including negotiating extended warranties, maintaining predetermined "fix-it" funds, municipal partnerships (e.g., collaborating with the municipal courts to bring in outstanding traffic warrants and the associated additional revenue), or dedicating budget line items to keep the system operational after grant funding or initial departmental funding expires.

Buying several systems "in bulk" can lower the unit cost. In some cases, jurisdictions funded by a single grant bought as a group a relatively large number of LPR systems.

All agencies noted that LPR systems require at least a minimal staffing level for sustained maintenance. For smaller LPR system deployments, staffing after an initial ramp-up period was usually much less than one full-time equivalent (FTE) staffer. There were added demands for staff time for system upgrades or work on underlying operating LPR systems. Agencies considered it to be significantly more work to implement the usually twice-daily hotlist updates and system data downloads by hand—using, for instance, USB sticks. As a result, most agencies elected vendor-provided options to transmit data and updates wirelessly.

Develop Policies for LPR Usage, Data Access, and Data Storage

Under the current legal and social environment, policies on this technology's use vary from location to location; local and state legal restrictions also vary. Agencies should develop their own policies for the use of LPRs that enable the types of uses the department will value while also conforming to local laws. Some departments have created policies more restrictive than local laws require, due to privacy concerns by police or local officials about aggregated data stored in a system. Those interviewed by RAND said that agencies just starting with LPR systems should work with departmental counsel to develop clear storage and retention policies for system data and PII if those guidelines do not already exist. They recommended that agencies also consult with local and state government officials about their privacy views to forestall problems, secure support, and alleviate concerns. Once agencies new to the technology develop use policies, they then should be prepared to revise them after learning from experience.

The policies, most importantly, must define who has access and clarify that system data may be used only for law-enforcement purposes. Most departments deemed it important to limit access and to define appropriate reasons for personnel to access the data/system. The system reads are imperfect, especially because plates are not standardized, so system-initiated alerts must be verified by humans, officers in equipped cars, those who can receive the notices in their vehicles, or, with fixed readers, dispatchers or others at a central, manned site. Departments typically created an authorized list of those with data access and gave them special logins and unique passwords. Agencies also crafted policies on purposes and maximum data retention times, as well as authorizations required to access stored information reads and "hits."

Think Outside the Box When Deciding How to Employ LPR

Agencies just starting out should "think creatively about how to use LPR data." While vendors often sell the LPR systems as a "stolen vehicle finder," police agencies interviewed by RAND found the technology useful in many other ways. They said they often learned from other agencies about new uses and that this sharing helped. One agency has suggested that LPR systems could pay for themselves if used a certain way—that is, if officers use LPR during overtime work to look for vehicles with outstanding traffic warrants. The revenue generated to the municipal court from this type of work could pay for the cost of LPR systems. It is similarly possible that a state can increase its revenue related to DMV violations (lapsed tags or inspections) if DMV data are pushed to local law-enforcement organizations.

Several departments store plate reads as a tool to help investigate crimes. By examining stored data for past matches to a suspect vehicle, investigators can get useful leads on a suspect's possible location. In addition, because LPR records include both license plate data and photographs of vehicles, eyewitness accounts of partial plates and vehicle descriptions can be combined to search LPR data for matches without having to refer to motor vehicle department records. The photographs in LPR records are also quite useful for responding to "be on the lookout" announcements. Agency representatives emphasized that a photo of the actual vehicle of interest taken by the LPR camera was much more useful than a general vehicle description (i.e., make, model, color, year) for spotting the vehicle on the road.

While "outside the box" thinking about this technology may, indeed, result in novel and effective applications, agencies also must factor in potential public responses. Suppose, for example, an agency decides to use a network of LPRs to monitor the vehicles of everyone who has been arrested for any crime. From a law-enforcement perspective, this might be an effective use of the capability, but it would potentially raise civil rights issues. Acceptable applications of LPR may vary from jurisdiction to jurisdiction based on prior practices, local laws, and political environment. Furthermore, more generally, new adopters may want to keep in mind that many LPR uses are likely to be challenged under the Fourth Amendment.

Enhance the Benefits of LPR via Cooperation with Other Agencies and Jurisdictions

Regional Cooperatives Enhance LPR's Utility

The ability to share LPR data among jurisdictions and agencies within a region enables law-enforcement agencies to use LPR information more efficiently and effectively. Vehicles suspected of criminal activity can be more easily tracked when each department within the cooperative has access to an entire region's database. Smaller departments can leverage the larger pools of information gathered by neighboring departments, and similarly exploit the information collected. Larger departments can work with smaller neighbors to track wanted vehicles. In addition, if a criminal uses a car to travel between jurisdictions to commit crimes, the information collected on the car in one jurisdiction can seamlessly inform investigations in another.

Police Departments Need to Link LPR Systems to "Effective" Data Sources

This technology's utility depends on the quality and integration of the IT infrastructure that enables it, feeds it information, and distributes its results. It can be crucial how well the system is set up to receive and share data. A key to system effectiveness is the type of data a system receives and how promptly and frequently it gets the information. LPR reads must be compared to hotlists held locally and updated frequently. Hotlists can be created for any of a number of types of situations or offenses associated with a license data—Amber Alerts, stolen cars, stolen tags, expired registrations, or emissions checks. An individual officer has a large amount of discretion in choosing whether to be notified about or to act on a given match, reducing concerns regarding potential overload. Users can toggle various types of alerts on or off to reduce the frequency of alerts and/or vary the severity of offenses that produce alerts. Both the sources and timeliness of hotlist data govern the ability of the LPR to do more than just archive data to answer future investigative questions.

Hotlist data generally fall into three categories: federally generated NCIC data (used by most if not all jurisdictions), locally added plates of interest or concern, and DMV-held data. Motor vehicle department data, in quantity and for their greatest impact on day-to-day patrols, may be the most significant potential information source, agencies interviewed by RAND said. Some jurisdictions did not or could not access DMV data. One jurisdiction said that without DMV data, it was using LPR systems to only "1%" of their potential. Many jurisdictions relied on federal data—in particular, the NCIC hotlist from the FBI. And some jurisdictions add their own license plates as a local hotlist. For clarity, Figure 7.1 depicts a hypothetical, large-scale, heavily integrated departmental LPR setup with the appropriate data coming in and out of the systems, as discussed.

The agencies interviewed by RAND said that LPR manufacturers do *not* typically set up information structure to push hotlists to users. Instead, IT experts at local agen-

Figure 7.1
Notional Large-Scale, Integrated Departmental LPR Setup

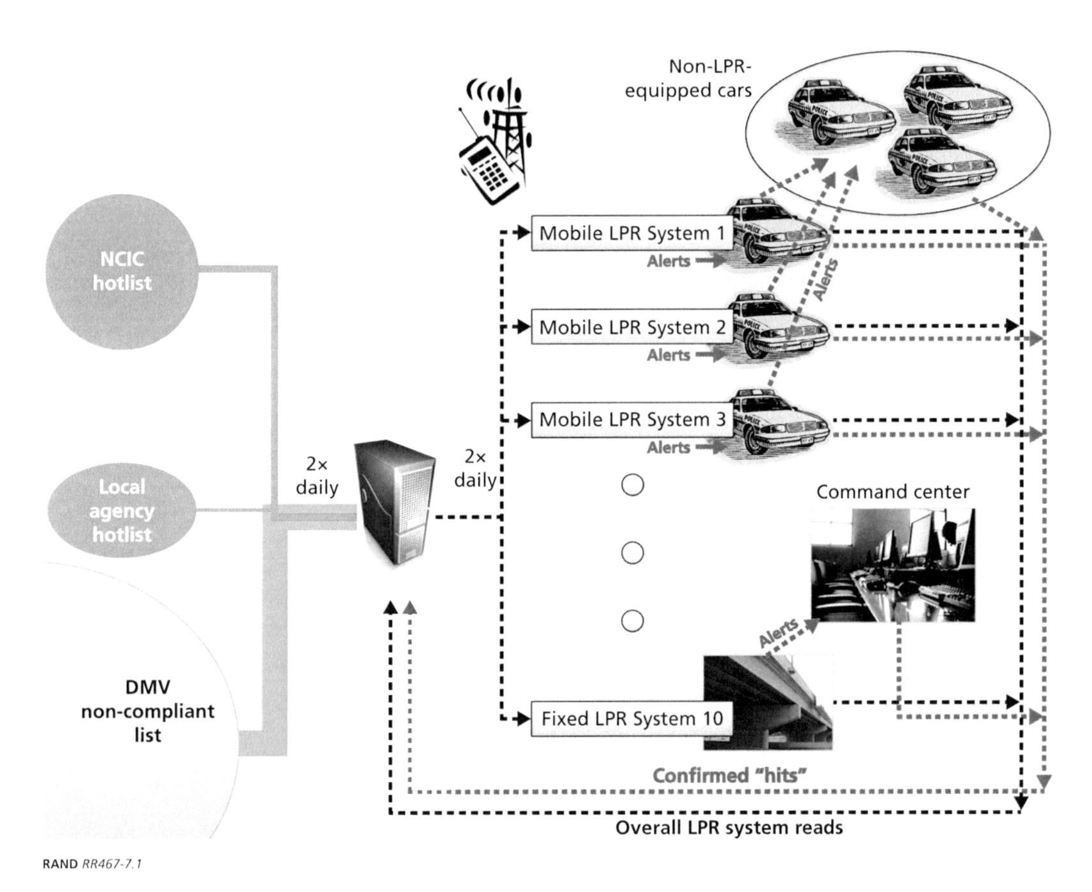

RAND *RR467-7.1*

cies coordinated—ad hoc and per agency—to create and update hotlists on various potential offenses or persons of interest. The basis for most, if not all, of these hotlists was the FBI-generated NCIC hotlist, which sufficed to track reported stolen cars. Local agencies will want to coordinate with state police or other jurisdictions to arrange for hotlist updates to be maintained and pushed out. Personnel in one locality interviewed noted that a local police department with exceptional IT talent had worked with the state to create the computer code necessary to perform this task; state police permission may be needed to access servers to update such hotlists.

Effective Implementation Frequently Requires Cooperation with Other Agencies

Agencies have found that sharing hotlist or LPR data with other law-enforcement jurisdictions frequently requires formalizing an agreement between the organizations.

Negotiating such MoUs with other agencies or jurisdictions can be both time-consuming and troublesome. To make things easier, agencies should consider appending LPR agreements to an existing MoU related to data-sharing rather than drafting a separate MoU specifically for LPR.

In addition to cooperation with organizations outside an agency's jurisdiction, implementing an LPR system often requires collaboration with other organizations within the same jurisdiction. For instance, if the city IT department supports the police, implementing LPR will require working with that IT department. Similarly, if another city agency owns and maintains police cars, it must be involved when police buy and operate vehicle-mounted LPR systems.

Fixed LPR systems can pose added interorganizational challenges, such as coordinating with relevant departments to secure locations, power sources, and maintenance service for LPR units. One department leveraged already installed speed cameras to co-locate its LPR systems, as power was installed and available at these sites.

Ensure "Successful" Implementation and Sustainment

Be Prepared and Committed for a Long Learning Curve

Agencies interviewed warned that new implementations of LPR can require a significant initial effort to set up, and that they in many cases had to change their plans or policies after learning more about LPR. Some departments, for example, started out by updating hotlists manually rather than wirelessly; having learned the effort involved, they then switched to wireless updates. One department determined the potential policing impact of LPRs by collecting data before and after LPR installation on patrol cars from individual officers assigned to them. The agencies almost all said they learned from other departments or professional associations. They characterized the effort to set up LPR systems as worthwhile; those agencies that ended up with the most automated and integrated systems subjectively seemed most satisfied with the result.

It Helps to Have an LPR "Champion"

Most, if not all of the agencies we interviewed could name a tech-savvy staffer who became a key to the success of their LPR system, providing crucial know-how about its purchase, setup, training, and maintenance, especially in the first year or two of operations. This specialist in some cases was not the system's sole champion but may have been joined by the police chief, who may have realized the new technology's potential, perhaps after hearing about it from other departments. The chief of police supported, as needed, the staffing, policy, interagency coordination, and financial needs of the LPR implementation. Details such as vendor vetting and integration with departmental IT systems usually were handled at the staff level by the technical expert.

Do Media Outreach to Facilitate Public Acceptance

One agency advised departments that were presently adopting the technology to "do a media blitz" to introduce local residents to it and its benefits. As recommended, engaging media gives the public that the department serves the opportunity to learn about the capability before it can be a surprise or otherwise pose any sort of problem. However, one other department recommended against this approach because it potentially could work against police efforts by "educating the criminals."

Plan to Maximize Use of LPR Units

The manner of assigning patrol cars that have LPR systems installed in them can play a role in the technology's success. Several agencies found important differences when plate reader units went into shared pool cars versus those taken home. Shared cars were in use around the clock, which provided the highest degree of usage of an LPR system. An LPR unit on a take-home car, however, was only used for one shift per day. In agencies that use both shared and take-home cars, therefore, it would generally be most efficient to install LPR units on the shared cars. However, some departments cautioned that if a less desirable portion of the fleet formed the shared pool, or if certain officers were uncomfortable with the new technology, LPR on shared cars would not be used as much as they otherwise would. They also recommended planning for how to keep LPR units in service while the cars they are mounted on go out for maintenance, especially because the system-equipped vehicles can rack up more mileage and need more care. A few agencies installed system mounts on more cars than they had cameras for, so they could shift these around when vehicles went in the shop.

Officer Training Is Important

Many agencies told RAND that officer training was crucial when installing LPR systems. This training included both officer training on how to use the systems and, separately, training for investigators on how to search the stored data. Vendors did not offer adequate training for detectives and others to effectively mine LPR data to the fullest potential, agencies said. Vendor training was characterized as limited and only providing procedural instruction about the technology itself, not addressing how officers should use and manage LPR data. One agency suggested that officers should not have access to the data without proper training. Another stated that officers seemed to learn more about the technology when a training course used case studies or stories that were interesting and salient. Many agreed that LPR training should be approached such that the technology is just another user interface/application that officers need to learn on the job. Finally, many agencies also suggested using a "train the trainer" method for LPR, allowing trained officers to teach others and, consequently, significantly increasing the exposure of new officers to trainers. Agencies generally conducted separate training programs for patrol officers and investigators.

Allow for Officer Discretion When Acting on LPR Hits

Officers should be granted the discretion to decide how to prioritize system alerts and hits within the rest of their workload. Individual LPR systems have toggle options to allow an officer to opt for which types of alerts he or she gets notification. Moreover, if officers get a notification, they also have the option to choose not to act on it because of other priorities. The notification toggles were especially important for jurisdictions successful in including DMV data, which resulted in frequent alerts if all possible alerts were toggled on. The option to have these frequent alerts was deemed an advantage because officers could always toggle alerts for the various types of DMV infractions on or off, as they deemed fit. One department indicated that data on stolen tags and registration increased the severity of offenses on which the patrol officers took action. One department demonstrated that it could show quantitative impact on the nature and quantity of officer activities by tracking daily data on all patrol officers (those with LPR systems and not) on arrests and traffic citations, over time, before and after installing the technology.

Plan for Data Storage

LPRs Generate Large Amounts of Data

The data files for LPR reads and matches are inherently quite large because they contain two types of images—both infrared and visible photos of the car and its license plate. Agencies repeatedly said they first underestimated their storage needs to keep data for given periods; these limits, rather than privacy concerns, ended up shortening their data retention period.

Cloud-Based Data Storage Lowers Costs but Runs Some Risks

Cloud-based data storage may ease agencies' retention problems and let them keep information longer if that is desired. This option can be very low cost, scale directly with departmental demand, and does not depend on an agency's infrastructure or require the same level of departmental IT support. It can also reduce the cost to acquire software licenses and facilitate mobile "smart phone" applications.

While cloud-based IT is gaining undeniable momentum in private and public-sector applications, it poses some risks, which most likely vary from vendor to vendor. The security arrangements for cloud storage of potentially sensitive LPR data (PII) could be inadequate or later could be perceived as such. Those interviewed did not cite it but another risk with cloud storage could be "vendor capture," wherein the high costs or impossibility of recovering data stored in this way with one company makes switching to another company difficult or infeasible. At least one agency said it had deemed cloud storage beneficial and worth the risk and had, in any case, kept its LPR data there.

As cloud technology matures over the next decade, it likely will become more generally apparent how much trust public agencies can put in the cloud service providers' security arrangements. A number of vendors currently claim to provide secure environments. Agencies considering this option need to look closely at the service-level agreement, data security arrangements, how data will be handled if the agreement ends (will it be deleted?), and recourses in the agreement for service failures or security breaches. It is possible that vendors of cloud services keep LPR data beyond times prescribed in departmental policies. Problems also could arise if the private companies were to use system data for their promotional purposes; this could expose the department to privacy complaints. Agencies must understand the exact parameters of how their cloud-based LPR data will be stored, for how long, and how it will be recovered and purged if a contract terminates. The potential liabilities for agencies in a shift to third-party cloud storage may be consequential.

CHAPTER EIGHT

Conclusions

While technology in its many current forms is improving the efficiency and effectiveness of police officers and agencies, LPR technology's potential to enhance police work is commanding. It can scan exponentially more plates and match them against hotlists for a range of infractions or individuals of interest. It can alert officers who, freed from the task of plate scanning and manual entries, may choose whether to act, working with greater information (including photos of targeted vehicles) than before. LPR systems also can allow investigators to analyze stored data for a wide variety of law-enforcement activities and investigations.

Because of its perceived utility, this technology has spread fast and wide in law-enforcement agencies since its U.S. introduction in 1998. Nineteen percent of agencies said in a 2007 survey that they used it. By 2012, 71 percent of survey respondents said they had it, and 85 percent said they planned to acquire it or expand their use of it. Although these responses came from different survey populations, the numbers reflect the keen interest by police forces to tap this technology.

RAND's research on LPR systems sought to go beyond usage statistics. We framed the technology's environment in reviews of literature and legal issues, then followed with exploratory interviews in seven case study agencies to find aspects of the systems' use heretofore unreported by professional organizations, academics, or the media. Our research uncovered and discussed unique elements of the challenges to this technology's effective deployment—factors such as cost, privacy concerns, resource allocation, personnel assignments, and the implications of data accuracy and standardization. We examined LPR systems' benefits such as specific force-multipliers in their technological versus manual identification and matching of license plate identification. To best summarize our examination of the range of ways in which this technology is used, its benefits, limits, emerging practices, and lessons learned, we conclude with key findings in four strategic areas: major use cases, privacy issues, interoperability issues, and recommendations for enhancing the utility of systems in the field.

Major Use Cases

LPR technology, initially promoted as a way to reduce car thefts, can be put to both reactive (real-time scanning, matching of plates to find stolen vehicles and targeting vehicle and traffic violations) and analytic (identifying suspects as well as crime trends and hot spots) uses. Existing research and our own case studies confirm that most law-enforcement agencies still use the technology primarily for auto theft investigations.

Agencies take less advantage of the LPR systems' analytic potential of using retrieving stored records for predictive policing or intelligence-led operations. Officers in these instances analyze stored plate records, which also include geographic and time data, to investigate drug- and gang-related activities, homicides, and burglaries. This kind of data analysis also has the potential to uncover trends in criminal activity that could help authorities target their crime-fighting money and resources. But because this approach relies on stored data, it inevitably prompts greater privacy concerns than LPR systems' reactive uses. This did not, however, appear to be the major hindrance to the technology's spreading popularity and adoption. While access to stored system data is an issue, so is the availability of resources and personnel to support the technology. It appears to be more useful for larger departments with greater resources and more personnel. They can also connect with neighboring departments to share information. Smaller departments see fewer ways and opportunities to take fullest advantage of the technology's potential boons.

LPR systems' utility is dependent on database access and data retention policies and practices. Agencies work with hotlists built with data from multiple sources; the more sources, the more effective the results are. But access to many and different data sources varied from agency to agency in our study; while most had access to NCIC or state law-enforcement files, access to DMV data was less uniform, and this can severely limited the utility of this technology. How long departments retain system data also proves critical to the analytical benefits the technology can offer. Departments that keep such data for a long time—months or years—can mine it for helpful information to investigate ongoing cases or identify crime trends and patterns. While we found that the categories of reactive and analytic were still relevant as a way to frame our research, the more important finding was that this technology can benefit any type of investigation, provided the necessary data are available to support the systems.

Privacy Issues

Without question, LPR technology raises privacy questions. LPR systems fall into a category of modern information technologies (such as the Internet and mobile phones) that have the potential to magnify individual uniqueness, thus raising privacy challenges. Concerns rarely involve aggregate checking of plates against hotlists. But law enforcement's storage of plate data for use in later investigations does raise serious concerns among privacy advocates. Such data could let authorities to reconstruct individuals' movements across space and time; if their patterns within the larger database are unique, other outside information can link plate data back to specific individuals. This can conflict with an individual's expectation of locational privacy.

The Fourth Amendment, which has at its core "the security of one's privacy against arbitrary intrusion of the police," would seem to address key concerns about plate reader systems. But advances in Fourth Amendment doctrine have fallen behind rapid progress in modern information technologies. Recent U.S. Supreme Court decisions on individuals' locational privacy have been conflicting or left key questions unresolved. One of the most recent Supreme Court decisions indicated that an individual lacks a reasonable expectation of privacy on open, public roads; another held that warrantless collection of location data over an extended period constitutes a search and, therefore, violates a person's expectation of privacy. The justices did not define what constitutes an "extended period of time." Uncertainties also reign as to exactly when law enforcement's use of plate reader systems likely violates the Fourth Amendment. Based on our research, we suggest that this technology's use has greater chance of being sustained under constitutional scrutiny if it can be framed as analogous to an officer's keeping an eye out for a specific car or plate associated with a particular crime; the more its use is seen as general surveillance, the more likely courts will find it problematic.

We regard as inevitable, particularly with the technology's widespread adoption and attendant increased publicity, Fourth Amendment–based lawsuits challenging its use. These cases may eventually reach the Supreme Court for prospective clarification of constitutional restrictions, if any, on LPR systems. Legal uncertainty, meantime, has led to some state high courts (i.e., New Hampshire, Maine, Virginia) to interpret their state constitutions as restricting this technology; privacy advocates there also have raised questions about data retention. It is important to highlight, however, that extensive system use is legal in most states and it is unlikely that any public users (other than in the states identified above) would face civil liability for their work with the technology.

Different agencies have adopted privacy policies for LPR systems' use, data access, and data storage. Authorities acted (or did not) with little information or guidance on how to incorporate privacy protections. They improvised as best they could. Technical or financial constraints, more often than policies, limit plate reader data storage. As media and published reports increase about this technology, policymakers will find it more important to put in place or improve policies for these systems.

Interoperability Issues

The utility of LPR systems depends on how well they are set up to receive and share data—both for hotlist databases and plate scans. Our research brought to the surface three challenges to interagency data sharing: the availability of the data; developing and signing administrative sharing agreements; and the technical interoperability of systems, including a lack of standardization of records. Access to data varies by agencies, and developing the necessary agreements to do so appears to be a major challenge; it takes resources even to pull together the needed technical, legal, and negotiation expertise. This means that agreements often are not put in place and data are not shared. Those agencies that want to share information also must use LPR systems purchased from the same vendor. Because two large vendors dominate the market, agencies in neighboring jurisdictions often operate similar LPR systems. This can only occur, however, if they know about this fact before they buy equipment and they coordinate their purchases. This may limit their choices and increase costs. Even with similar systems, data often cannot be shared because of differences in type, format, and quantity of information in records.

While these three challenges may appear the most problematic, they also potentially have simple solutions: sharing hotlist or LPR data among law-enforcement jurisdictions often requires new, formal agreements, and developing such agreements has created a considerable barrier. Instead, agencies should forgo separate materials and consider appending their agreements about these systems to existing MoUs on data-sharing. As for vehicle scans, these must be compared to hotlists held locally on the system. These hotlists generally fall into one of three categories: federally generated NCIC data, locally added plates of interest or concern, and DMV-held data. Because the idea of sharing this information across LPR systems' databases is relatively new, the data structures often are incompatible. National standards should be developed and disseminated for these databases and data records to greatly enhance information-sharing across agencies, and to boost the utility of plate reader systems.

Recommendations for Enhanced Use of LPR

Besides key findings in the four strategic areas outlined above, our research uncovered key principles to enhance this technology's use among law-enforcement agencies. Departments considering LPR systems, before they make big investments, may wish to heed these principles. They should do the following:

- Estimate and secure necessary funding for the entire lifecycle of the technology. This technology-based equipment must be not only purchased but also updated and repaired. The equipment is of no use if sidelined by maintenance issues. Funding should also take into account costs for minimal staffing to operate and maintain LPR systems as well to train those using the technology at the beginning and later on.

- Ensure that sufficient IT infrastructure is in place to handle different types of data promptly and frequently. It is both the sources of data (Amber Alerts, stolen cars, stolen tags, DMV data, etc.) and their timeliness that determine if LPR systems are used to their fullest. The structure must be able to handle data coming in and going out of the system so departments can effectively cooperate with other agencies and jurisdictions.
- Develop policies for system data use, access, and storage as soon as possible. Clear guidelines on use, storage, and retention of data will ensure that personnel consistently and legally deploy the technology in ways that bring the most value. Agencies lacking policies may wish to consult departmental counsel and local and state officials to forestall problems. They should be prepared to revisit and revise policies as public or political support and/or concern over the technology's use may change over time. Agreements may need to be put in place to share data with law-enforcement partners (both inside and outside of an agency's jurisdiction). This requires negotiating MoUs for all types of data-sharing, not just for LPR systems.
- Think strategically about integrating the systems into the agency's day-to-day operations. Departments should expect and be prepared for significant initial effort to set up their systems, as well as a long learning curve to see their full use. The process can benefit from the presence of a tech-savvy champion and leadership from the top.
- Consider this technology as more than a "stolen vehicle finder." To improve its cost-benefit ratio, all opportunities for its use should be considered. Agencies should learn from other departments how to effectively and efficiently expand their use to more analytical applications.

Here are some practical next steps from RAND's research to facilitate more effective and less problematic adoption of this technology by law enforcement:

- Our research highlighted the ways in which this technology has the potential to transform both reactive and analytic policing, particularly if advocates and agencies will address two relatively easy to overcome challenges by developing model MoUs for agencies to share plate reader data so as to enhance LPR systems' uses in analysis and by developing model privacy policies for these systems, with configurable sections depending on use cases.[1]

---

[1] Several models for MoUs do exist. The IACP website provides an example from the Charlotte-Mecklenburg Police Department regarding an MoU between the CMPD, Mecklenburg County Area Mental Health Authority, and Mecklenburg County Department of Social Services for the purpose of cooperating in the Child Development-Community Policing initiative. Although they are not LPR-specific, these templates provide

- While law-enforcement personnel see this technology as a positive development, privacy rights advocates have raised issues that might curtail its future potential, particularly if agencies are uncertain about particular use cases. Risk taxonomy technical standards can be developed based on information in this report to identify and estimate exposure to privacy risks and specify potential law-enforcement utility and privacy rights "tradeoffs."
- Data gathered in this research also could be expanded to produce an online tool for cost-benefit analyses of LPR system use (considering costs of acquisition, implementation, and maintenance) vis-à-vis crime deterrence. Police agencies at all jurisdictional levels could tap such a tool to counter advocacy, political, and public concerns about this technology's use in criminal investigations. The calculator could be developed so it uses local information (or relevant population and crime statistics at whatever jurisdictional levels) to customize the benefits.

To expand LPR systems further into careful, considered, intelligence-led policing, such as by using analysis of data from historic databases to identify crime patterns and trends, LPR proponents clearly must build and spread needed information and tools. This would empower agencies to deploy the technology so that it brings the most value in these times of tight budgets, allowing law enforcement to target its resources on operations with high gains.

---

a basis for developing a specific language for cooperation between agencies. See Charlotte-Mecklenburg Child Development-Community Policing MoU, June 2006.

APPENDIX A

Case Study Summaries

In this appendix, we present summaries of the interview information gathered from the seven case study law-enforcement agencies. To encourage frank discussion of issues, we anonymized participating agencies and do not identify them.

DMV-Enabled Cluster County

Information About the Agency

DMV-Enabled Cluster County has a population of roughly 1 million. It borders both Cluster City and Cluster County and has a significantly higher per capita income than the sample median. It is a relatively wealthy and low-crime area.

Use Cases

While many agencies' primary motive in acquiring LPR was to help recover stolen cars, the systems in practice can be used for a range of purposes: tactical (dealing with vehicle, traffic, or parking infractions) and investigative (retrieving stored records in targeted crimes for suspects or determining trends or hot spots). These two uses by far account for the most use of the system here. The agency maintains hotlists for stolen cars, terrorist activity, and homeland security issues from the NCIC. Alerts come from the Vehicle Emissions Inspection Program; little investigative or other analytical work has been done with its system so far.

Implementation

The agency initially acquired two units with state funding, then got about 20 more as part of a regional federal Homeland Security grant covering 32 agencies. It took about a year for technology startup, including troubleshooting with the supplier. The system is managed by a lieutenant and requires less than one FTE to manage; this individual is very knowledgeable about the system, but the agency relies heavily on vendor support for software troubleshooting.

The agency operates 22 mobile units (out of a fleet of 800 cars) and one portable (trailer-mounted) unit. It has no fixed units but has plans to obtain some; a regional transportation authority operates a fixed unit, but the systems are entirely separate.

There are two training courses, each two hours long: patrol officers learn about the software and how it works; others may be trained on server queries—how to make them, what wild cards to use, and how to compare commonalities among queries. This training's most important aspect is teaching proper control of the data under agency policies. Data are for official law-enforcement use only, not for personal use or for use by another agency/department. Server data can be accessed by investigators and civilian analysts.

National and state criminal hotlist data (from the FBI's NCIC) are updated to LPR units twice daily. State motor vehicle department data (registration information, suspended driver licenses, vehicle emissions compliance) are updated once a day. The system automatically merges the separate data streams into a single database. The agency lacks but is working on a policy on system use.

The system gets more hits against motor vehicle department hotlists than officers can respond to—in some situations, units record multiple hits per minute. This means that officers must exercise discretion in choosing which hits to act on. The system lets users select the classes of hits for which to send out alerts. If officers get bombarded by too many alerts, they can deactivate alerts for certain types; they sometimes deactivate alerts for low-level hits so they can stay focused on more-serious offenses. Before acting on an alert, an officer must verify it with the dispatcher or have probable cause.

System data (scans and hits) are stored for one year; privacy was a factor in choosing this duration. The agency chief is a proponent of cold-case investigations and has developed a policy on archiving certain hits indefinitely for this work. Data for this purpose would get stored on media physically disconnected from the network and locked in a limited access room. Access to this archived data would require permission from the chief or assistant chief. However, the agency lacks the hardware or resources needed to implement this plan.

Some plate reader units are on shared cars, and some are on take-home cars; there are pros and cons for both. Take-home cars operate only one shift per day, so the units here are underutilized. The shared cars, meantime, are old and dirty and are not get used often either. The agency plans to move its units to new shared cars to increase their use.

Interoperability/Information-Sharing Cost Issues

This agency does not yet share with other police agency servers because it still needs to develop an MoU; so although the technology is in place, administrative aspects are unfinished. When systems upload data from cars to the agency server, they also automatically upload to a statewide server, and information from other agencies in the state is also copied there. Data from other agencies can be requested from this state server but only manually.

Other Issues/Findings

- Implementation was burdensome. This system can create a big information overload for the officers.
- DMV data can cause an excessive hit rate, creating an information overload and stress for officers. Managing and prioritizing alerts is key.
- Officers wish they could restrict certain alerts to specified locations or designated times; they do not necessarily want an alarm for every hit about a sex offender but might want it if the match is near a school. This capability is unavailable now.
- Because security concerns generally do not allow law-enforcement agencies to provide remote access to their computers, system troubleshooting by vendors can be more difficult since they must do this by traveling in person to agencies.
- This agency occasionally has problems with hotlist data updates. New files with new date stamps were received from the state twice daily, even though no updates actually occurred. At one point, the data were three weeks out of date and accordingly were problematic. The system administrator ended up writing computer code to confirm that newly received files differed from prior versions.
- Drivers have beaten the system by using black electrical tape to alter their license plates.
- While the agency has not done a comprehensive analysis of the system's effects on outcomes, preliminary analyses indicate the technology has increased traffic officers' productivity, allowing them to identify the nature of an alleged violation before making a stop; the system lets them concentrate stops on more-serious violations. Reviews of individual officers' stop records, before and after the system was installed, show roughly a tenfold increase in jailable offenses (e.g., hit-and-runs, DUI, driving without insurance, failure to appear in court). The total number of stops did not change significantly, but officers were forgoing less-serious violations to concentrate on more-serious ones.

Cluster City

Information About the Agency

This agency, selected as a cluster city, is characterized as above the median of all agencies in the sample in the size of its force and the size and income of the population it serves. It also is above the median in terms of crime rates.

Use Cases

The agency primary uses LPR technology to recover stolen cars and stolen license plates, which accounts for 90 percent of the system use across the agency. The technology helps officers recover stolen cars more quickly; because they typically are abandoned soon after they are taken, this means that recovered cars suffer less vandalism and damage. The system also helps to apprehend fugitives, to locate vehicles observed at crime scenes, and to monitor known criminals and gang members.

Data for the system's hotlist are provided through the FBI's NCIC and supplemented by local activity information added by the agency, which lacks access to state motor vehicle department data in its database. It does not, as a result, use its system for such vehicle violations as revoked, suspended, or expired driver licenses or vehicle registrations.

The jurisdiction's public works department uses this technology in residential parking enforcement, to "boot" parking scofflaws (vehicles with more than two outstanding parking tickets), and to identify out-of-state vehicles that need to be registered locally. The public works units alert police about hits on vehicles associated with police matters (e.g., stolen cars or wanted persons).

Implementation

This agency acquired most of its LPR units under a large federal Homeland Security grant covering 32 agencies. It previously had an early, vehicle-based, non-networked system strictly used for stolen cars. The current technology's management is led by a civilian employee with experience in data system management. The system requires two or so FTE positions to manage.

The agency has plate readers on 25–30 cars in a fleet of about 1,500. In each of seven patrol districts, there are two to three car-mounted units, and the agency operates 38 fixed units installed around the jurisdiction. Those fixed units were installed at sites also used for other technology, such as traffic- and red-light cameras. Although similar in number, the fixed units record the most reads, compared with their mobile counterparts; mobile units generate roughly 1 million reads monthly and fixed units record the rest of the 16 million to 17 million monthly reads for the agency, which also operates one portable (trailer-mounted) unit.

Given the department's size and relatively small number of units, not all officers are trained to use the system. The agency picks officers for its system-equipped cars and trains them to use them. Officers are chosen partly based on their interest in and willingness to use this technology. Patrol officers undergo a four-hour course on the technology's hardware and software taught by the civilian system manager and an officer experienced in its operation.

Detectives and other investigators who may perform server queries get more informal, one-on-one training with an experienced user. All detectives can be granted access to server data if they need it. A general order requires users to have a law-enforcement need to access system data.

The fixed units require more attention and resources to operate than do mobile units. The agency's information center monitors the fixed units 24–7, and staff there inspects all alerts and coordinate with dispatch to assign patrol car to investigate when appropriate. The department must coordinate with other agencies and property owners to install its fixed units, which require permissions to mount and access to electrical power (this is a minimal cost, but a bill for which payment arrangements must be made). Fixed units are exposed to varied weather and the agency must install and maintain them, which can require access by a bucket truck. As with mobile units, fixed units require high bandwidth wireless data transfer capability.

Hotlist databases for mobile and fixed units update wirelessly twice daily. All system data (reads and alerts) are stored for two years, and there are no time limits on keeping information that is part of an active investigation and that may be saved as part of a case file.

Interoperability/Information-Sharing Cost Issues

This agency does not share data with the motor vehicle department because it feels it has sufficient police work to pursue without having this information, too. It also lacks direct access to other law-enforcement agencies' servers; MoUs are in place with some agencies to facilitate sharing, but access to data from others still requires calling to ask them to manually query their servers. Sharing agreements are in the works with more agencies.

The agency shares data with the public works department in its jurisdiction for parking enforcement and locating stolen cars. It can share "hits" from mobile and fixed units with police cars not system-equipped to assist in responses when needed.

Other Issues/Findings

- The agency wants to measure its system's effect on outcomes and the chief's daily summary carries system-related arrest data. But the agency has not begun to systematically develop statistical measures of its system's use. Its results are compiled and communicated manually.
- Limited storage for system information poses problems and the agency is not saving high-resolution data because of the file size. Even so, it is running out of space and is upgrading to a server with more capacity.
- Storage limits how long system data can be kept. While privacy advocates may call for retention limits, the agency noted a federal regulation requiring authorities to keep for three years PII that may be legally discovered; such a retention time, however, was deemed infeasible because of data storage limits.
- Wireless communication speed also imposes limits, with some transfers of system data from units taking hours. The agency is upgrading to a 4G wireless system and is considering hardwired fiber links for fixed units.
- Grant funding hampers the agency in repairing and upgrading its system because administrators must go through a grant manager every time they need funds.
- It is key to have a clear general order, such as a department policy, in place before deploying system equipment.

- This agency developed its system use policy based on a sample provided by the U.S. Department of Justice.
- MoUs are difficult for police departments to negotiate, and they may wish to append language about their plate readers to existing data sharing agreements with other departments, rather than drafting separate accords addressing just this technology.

Cluster County

Information About the Agency

This agency, chosen for the cluster category, is above the median of all departments in the sample in the size of its force, at the median for the size of population it serves, and below the median in crime rates.

Use Cases

This agency primarily uses its system to recover stolen cars, although it also applies it in other ways, too, including to locate wanted persons, such as suspects in an active case; to find those with outstanding warrants; to place those on terrorist watch lists, subjects of "be on the lookout" dispatches (including Amber Alerts), and "deadbeat dads." It installs units at target sites, such as special events that could attract terrorism suspects and crime hot spots. It has searched archived data to help identify possible suspects in hit-and-run cases.

Data for its system hotlist come from the FBI's NCIC and a state-level crime information system. This agency lacks access to state DMV information in its system database, so it does not use this technology for such vehicle violations as revoked, suspended, or expired driver licenses or vehicle registrations.

Implementation

This agency acquired its plate reader technology four years ago under a large federal Homeland Security grant covering 32 agencies and roughly 250 system units. The agency led the initial proposal and still manages the grant for all participants. The agency sought this technology after learning of its benefits while working with a federal agency that used a system for stolen vehicle recovery. Its car-mounted units cost roughly $17,000 each, bought in bulk.

The agency operates 10 units: nine mobile and one on a portable trailer. It typically deploys mobile units on three to four of 15 or so patrol cars on each shift. Even as it plans to expand its network by acquiring fixed and more portable units, it is seeing if can expand its system to parking enforcement vehicles.

The system is managed primarily by a criminal investigations unit detective and an auto theft unit lieutenant. Officers in patrol, investigations, and auto theft use the system, which requires roughly one FTE position to manage (including training, IT support, and vehicle installations).

Patrol officers receive one to two hours of training from a system manager; it includes a slide presentation and vehicle demonstrations. Investigators and auto theft officers who must access server data undergo three to four hours of supervised training, plus many hours of self-guided, on-the-job training. Those who access the server must register and receive a unique username and password. Fewer than 10 percent of sworn officers in the agency get server access, a percentage kept deliberately low to minimize the risk of data breaches or improper use.

Data are transferred to and from system units wirelessly via a commercial mobile data network twice per day. All system reads are stored for six months and all reads that match a database entry (an "alert" or "hit") are stored for 24 months.

System-equipped cars typically are deployed in relatively higher crime areas. All system alerts must be verified with dispatch to ensure they are still valid before an officer may act on them.

Interoperability/Information-Sharing Cost Issues

Data sharing is a challenge for this agency, which now shares with only a few other departments that have direct access to each other's servers. Data must be requested and manually shared with all other agencies, meaning that an officer must call another department to ask it to run a given server query. This can take time; it requires personal relationships among staff.

The agency lacked a server when it first installed plate readers, so its personnel had to move data manually to and from system units using a USB drive, as well as logging all hits manually. This required too much work, and a server soon was installed.

The agency's equipment and data from its vendor are incompatible with those of other departments that use a different manufacturer; the systems are not interoperable. However, all agencies participating in its regional grant use the same manufacturer.

The agency has declined, thus far, to share data with federal agencies because it wants to avoid involvement in potential litigation related to tracking non-suspects.

Other Issues/Findings

- This agency has not analyzed the impact of its system on the recovery of stolen cars or the identification of wanted suspects, but officials were emphatic that they valued the technology because it has saved thousands of work hours. This agency recommends that recruits at the police academy undergo system training so that individual departments need not devote valuable resources for this purpose.

Small Town

Information About the Agency

This small agency was selected in the Big v. Small category. It is below the median of all agencies in the sample in size of its force and the size and income of its population served; it is at or above the median in crime rates.

Use Cases

This agency primarily uses its system to recover stolen license plates and stolen cars, to make investigation-specific queries of archived data, and to identify people on federal terrorism watch lists. Officials praised the system for also providing archived photos of wanted vehicles to help confirm the accuracy of its vehicle alerts.

Data for the system hotlist come through the FBI's NCIC and the state-level crime information system. This agency lacks access to state DMV information in its database, so it does not use its system for vehicle violations, such as revoked, suspended, or expired driver licenses or vehicle registrations.

Implementation

The agency first acquired plate readers as part of a regional, federal Homeland Security grant, which paid for two car-mounted units at roughly $20,000 each, a price that includes an extra $2,000 to $3,000 for a third, rear-facing camera for each unit. These rear-facing cameras were added because this agency is near a border with a state that does not require front license plates. The agency hopes to acquire some covert, portable units for surveillance use.

Its system is managed by a designated patrol officer. Installing the system was labor-intensive, but it now requires little time (much less than one FTE) to manage. The agency has a vendor support contract with an annual cost of roughly 10 percent of the initial capital investment.

Patrol officer training is informal, featuring a vendor-supplied slide presentation and a hands-on demonstration. Training for server use also is informal, consisting primarily of on-the-job self-training. Server access requires users to register with a unique login and password.

Data are transferred to and from system units wirelessly via a commercial mobile data network twice per day. All system reads, including those matching a database entry (an "alert" or "hit"), are stored for six months, a duration driven primarily by data storage constraints. Privacy is a secondary concern.

The agency puts its system units in higher-crime areas of the jurisdiction. It gets only a few ("a handful") of alerts per month. The system can send alerts to all cars on patrol (those with or without units) but this option has not been used so far.

Interoperability/Information-Sharing Cost Issues

This agency cannot share server data directly with any other departments. It occasionally gets requests from neighboring agencies, and it queries them for archived plate reader data. It wants to share server data directly, but this requires more resources than the agency has available.

Other Issues/Findings

- The agency has not analyzed its system's impact on vehicle recoveries or in other areas—but, qualitatively, it has had a positive experience with this technology.
- Cost is an important limiting factor for small departments in acquiring, maintaining, and fully utilizing plate reader systems, the agency noted.
- Installing the system can be labor-intensive, but operation soon becomes relatively simple.
- It is important to develop clear policies and practices on system data use and to revise those as appropriate.
- Agencies must consider the need to coordinate with partners (e.g., the municipal IT department or vehicle pool) about buying and running this system.
- This agency chose to do a high-profile "media blitz" to minimize the mystery and apprehension about what plate reader systems can and cannot do.

Big City

Information About the Agency

This case study covers a large police department in a city with a medium-sized population (between 200,000 and 400,000) and below-average per capita income. The city's property and violent crime rates and vehicle theft rates were all higher than the median for all departments used in the sampling methodology.

The department, headquarters for a large plate-reader regional cooperative with 84 agencies in it, itself has four fixed and 109 mobile plate-reader units. The regional cooperative (including the department) has a total of 134 mobile and 14 fixed units. Twenty-six mobile units were self-funded; the rest of the system was purchased under state and federal Homeland Security grants. The first units were purchased in 2007, seven more were acquired in 2008, and the rest were bought in the past five years. The last federal Homeland Security grant was awarded in 2011, so the plate reader funding will be depleted in 2014. Besides overall system maintenance costs, the agency must pay to maintain the system software and cloud-based database, as well as to cover its contribution for the regional headquarters operations (which share space with the department). The regional cooperative provides significant benefit to small participating agencies; they can leverage operation, maintenance, and management of the larger network, and the only costs they incur are those associated with switching out system units mounted on vehicles going out of service for maintenance.

Use Cases

While this technology was targeted at first for recovering stolen vehicles, the system's greatest benefit to the agency stems from analytic operations. This region has six dedicated analysts who mine plate reader data, mostly for investigation-based purposes. The department does not do much overarching analysis because it lacks the necessary analysts. The agency said that if it had more resources (i.e., more analysts) it would undertake more intelligence-led analysis. The department values its plate reader system because of its "back-end" ability to mine data to assist ongoing investigations.

The department mostly uses its fixed cameras in a reactive manner. Alerts come in to any of the database users, but many times no one is watching for hits on the other end. Fixed cameras for this reason seem to be mostly used by smaller agencies with more time to investigate hits.

The department specifically mentioned its uses of this technology include finding missing persons and solving hit-and-run accidents. The system also makes it relatively simple to cross-reference license plates from a sexual offender database with a description of a vehicle and with a time stamp.

Implementation

The informal and personal relationships formed in this region among officers and administrators facilitate more effective system operations. The regional coordinator is the liaison between the agencies and system vendors; if problems crop up, the officers call the regional coordinator, who can fix the problem on the spot or can coordinate with the vendor to solve the problem. This type of relationship has been necessary as many agencies experienced problems with the vendor's customer service.

This agency considers it important to train on how to use the system data, as well as on how officers on patrol should approach a motorist whose vehicle comes up on a system hit; departmental policy says that every hit must be verified, and this, in turn, leads to no "false reads" on the system. Even if a system unit indicates a hit, the officer must verify it. The system vendor gave agencies in the region no training on how to use or analyze data, and the regional coordinator now spearheads a "train the trainer" program.

Interoperability/Information-Sharing Cost Issues

With 84 participating agencies, the regional center has 1.5 terabits of data storage on a cloud-based system and more than 200 unique users of the system database. Every agency has the same vendor, which makes interoperability possible. Even if one participating agency lacks a plate reader system, it still can access the system database. The department runs a fusion center with six dedicated analysts who have access to video surveillance and plate reader databases. Other agencies can ask the fusion center for help with analytic information. System data are kept for one year. System software allows officers to complete searches both at police stations as well as in their cars.

The region has access to a number of hotlists, including the NCIC, a state-based hotlist similar to the NCIC's, and a more descriptive regional hotlist database updated every few hours. This last hotlist may include not just a license plate hit but also a description of the car owner's crime. All data in the hotlists come from vetted sources with established procedures for entry, removal, and audit.

The system's database software allows for an infinite number of users; the regional coordinator insists that anyone with access must be trained, and this keeps officers outside the regional co-op from access. As deterrence to accessing system records for unofficial reasons, the software logs all searches by the user's unique identification.

Other Issues/Findings

- System-equipped cars rack up more mileage and do so much more quickly, resulting in more frequent maintenance and putting vehicles and their mounted mobile plate reader units out of commission. The cost to switch system units among cars in service cannot be justified; instead, system-equipped vehicles can be out of commission often.

South City

Information About the Agency

This case study describes a large police department in a city near the U.S. border with a medium-to-large population (between 500,000 and 750,000) and below-average per capita income. The city's property and violent crime rates and vehicle theft rates were comparable to the median values for all departments used in the sampling methodology.

The department got its systems in 2009 with funding from state grants for border security. The agency bought one plate reader unit with the grant, found it beneficial, and then sought more: units for 12 cars, eight trailers, and three portable systems. The purchase price included a vendor system warranty for three years, coverage that had expired at the time of RAND's interview. Because its systems need upgrades, the department has applied for new grants but has been unsuccessful and cannot expand this technology until new funding can be found.

Use Cases

The agency mainly uses its system for auto theft, not necessarily in prevention but in recovery of stolen vehicles, which is important because it often leads the department's investigators to target drug cartel operations. The department said it has identified nine drug-trafficking organizations using its plate reader system.

That technology also is used in analytic investigations, in which officers, for example, will drive a system-equipped car to or park a system-equipped trailer in an area of interest to obtain revelatory data. Probable cause, as in an open criminal investigation or case, is needed, however, before investigators can search the system database and analyze information from it.

The system's managing sergeant is spearheading a novel idea to partner with the municipal courts, seeing if the courts might pay officers to tap this technology to find offenders when system hotlists get heavy with traffic warrants. The fines violators pay could be split with the court and the department could cover not only the costs to track offenders down but also to repair and expand the plate reader system, giving it even more capabilities. The department has pitched this idea to the city, which seems receptive and has taken it under consideration.

Finally, this agency said the plate technology improves its officers' safety, giving them key information before they stop and deal with motorists, about whom they now know much more. Overall, this department said it has only tapped the surface of this technology's benefits and there are many other case uses to be explored.

Implementation

This agency is experiencing severe difficulties with its system's post-warranty maintenance costs, such as its trailer units becoming inoperable because of issues in the generators powering them, as the RAND team saw on its visit. The managing sergeant, who noted the lack of repair manuals, said he learned to repair the systems by watching vendors. Absent proper maintenance training by the vendor, system administrators must learn on the job, and repairs are time-consuming. With its accumulated knowledge, this agency helps other departments troubleshoot their systems.

This department employs a "train the trainer" program for officers operating system-equipped cars. The agency struggles to get more senior officers to use the plate reader system whose units often are mounted in the newest cars, which are assigned to them. The department hopes to buy portable systems that can be checked out by more-junior, IT-savvy officers who want to use them.

The agency stores its system data on a server it bought, and retention time is limited only by its memory capacity. The department said it eventually wants to store data for a year and is looking at cloud-based systems to do so.

Interoperability/Information-Sharing Cost Issues

The department operates its system databases out of a larger fusion center, which gathers and analyzes media, including video surveillance, from sources such as school districts and the state National Guard. Other agencies (including the Guard and Child Protective Services) have access to this data because the department has set up MoUs with each of them. To access the system data, however, agencies must request this from the fusion center, based on an open criminal investigation; retrieval of stored records and analysis occurs on a specific case basis. If data get pulled as evidence in a case, they can be stored longer than is possible on the department's system server. The department itself has only five people with access to system data without case-specific need.

The agency uses a wireless connection between its plate reader system and server hotlists. It has access to only the NCIC hotlist and state version of the NCIC. These list stolen vehicles, Amber Alerts, and missing persons. The department hopes soon to access warrants and criminal records. As with other agencies, it lacks access to DMV data because it would be charged for it. Vehicle units receive updates with a new hotlist every 15 minutes; the state's version of the NCIC updates every hour.

The department has an inefficient relationship with the federal border security agency, which provides data to NICB; although the department has access to these data, the NICB data are not part of its system hotlist. Similarly, the department has access to the border country's plate reader data only by request, and their relationship on such data sharing is informal.

The department envisions a regional system that shares system data through the already-established MoUs for the fusion center. A small town outside of the department's city has one plate reader unit and provides data to the city database. The town can request data for specific investigations, though it lacks back-end access. The department also may share with college police who have systems (incompatible), and repossession companies.

This department has not experienced privacy concerns about its system or its data. The fusion center has processes to safeguard data, and one information request to the department from the ACLU was handled relatively seamlessly. The department has no plans to buy fixed units, which may create greater privacy concerns.

Other Issues/Findings

- This department makes no secret of its system but does not publicize its use, considering it "just another tool" and not wanting "criminals" to learn how officers got on to them so they could defeat this technology.
- Senior leadership have bought into the system significantly; the RAND team spoke to the assistant chief, who called the system an effective tool.
- This department said the system's value is linked to the number of data sources (hotlists) available.

North City

Information About the Agency

This case study covers a small police department located in a city near the U.S. border with a relatively small population of fewer than 100,000. The city has below-average per capita income and high property crime rates. Its violent crime rates and vehicle theft rates are comparable to the median values for all departments used in the sampling methodology.

This department acquired its first LPR unit in 2007 and now has a total of three units, with five more units on order. As with many other agencies interviewed by RAND, the department said it got its first unit under Homeland Security grants from its state's Department of Criminal Justice Services. The state chose this department and asked it to apply for a grant because it has potential terrorism targets in its jurisdiction. Based on its experience with the technology, the systems administrator encouraged the department to seek funds for more units, and a federal grant paid for two more. Two of its three units are fixed on patrol cars and one is mobile and can be transferred among vehicles.

Use Cases

This department sees its greatest system benefit in traffic enforcement. Under its original grant agreement, the department had to keep system-related statistics and found that the number of arrests generated from normal traffic stops increased dramatically when the technology was installed. The system's administrator said police had not increased the number of stolen vehicles recovered with the system but they had been effective in finding vehicles with suspended registrations, lapsed insurance, and parking infractions. Since these generate revenue, "LPR pays for itself," the administrator said.

Although this department serves a border city, it has minimal contact with the U.S. Border Patrol in regard to its plate reader system. The patrol has fixed units at the border but its data normally do not get shared with this department, making its U.S. border-related use cases rare.

The department occasionally uses system data for analytic purposes, recalling records from the databases only when specifically called for by an investigation. Other department divisions may provide a plate number to the system administrator, who is a part of the traffic division and is the main point of access to the data. The typical analytic use by this department is to help identify suspects, as opposed to locating or building cases against suspects identified by some other means. The administrator has discussed having the system map locations where the database finds a license plate of interest and highlighted as a major benefit the technology's ability to find matches with only partial plate readings.

The department has used its system in counterterrorism activities. Because its system grants were for homeland security purposes, the department was required at one point to drive its LPR-equipped vehicles to an industrial facility identified as a possible target of terrorism. The department also has assisted the FBI in investigations in which federal authorities directed the police, without explanation, to deploy their system-equipped vehicles in specified areas.

Overall, this department said its system's main benefit is its ability to increase the efficiency of activities in any of the case-uses described above. They described it as "just another tool" that lets them tackle many activities with minimal effort.

The department hopes to adopt a cloud-based database, which could help it coordinate with Native American reservations in the area, resulting in more vehicle-theft investigations. The administrator said the existing system would be used more and new units would be added, if the department had the funding.

Implementation

It is worth noting that this department's two vehicle-mounted plate readers (it has another mobile unit now and five more on order) have two cameras, one on each side, and the cameras angle is controlled by the vehicle operator.

The department said it has experienced no maintenance issues with its system; its vendor identified and fixed an issue before the department knew it was a problem.

This agency lacks a central database to keep system reads and the vehicle-mounted units actually hold the system data. So if the department wants to search the system database for a plate, it must run that search separately on all three units it owns. The current database can search on partial plates. The lack of a central database also means the department downloads its hotlists to each car separately.

It gets its hotlists from the state DMV, which sends its version of the NCIC hotlist, as well as out-of-state or border country data reported to the state. The hotlist includes suspended registrations, missing persons, stolen vehicle, and warrants. Unlike other case studies, this state DMV provides data at no charge to the department, which also can upload its own hotlists into the system database (e.g., information on Amber Alerts, probations, and burglaries). The department decided to use DMV information that included NCIC data instead of obtaining that federal information separately. That is because its plate reader cameras incorrectly identified certain out-of-state/country plates. The cameras cannot differentiate plates from different states and countries. Because its area has many tourists, the department found the number of incorrect hits was becoming an inconvenience.

When the department receives has its five new units, it will switch to cloud-based storage for its database and it has bought conversion kits so its existing units can access the new system.

Interoperability/Information-Sharing Cost Issues

This department has a high level of control over who sees and uses its system data. Generally, only system administrators have access, and data are held for 90 days or until the vehicle databases run out of space.

When the department moves to a cloud-based system, any of its personnel can be granted access by the system administrator, and a mobile application for database searches also soon will be available. The department will offer different levels of access to the database—for system administrators, for IT personnel, and for detectives.

The new cloud database can store data indefinitely. This technology's vendor also contracts with private-sector clients in banking and car insurance. This means that the department will get access to data collected by these private businesses. Its agreement bars its vendor from sharing the department's plate reads with private clients. That five-year agreement, if not renewed by the department, gives the vendor rights to own the police plate data, which it may not sell, though it can be used by other law-enforcement agencies. The data will never be made available to private clients.

The department's sharing of data with the border patrol has been rare, occurring on an as-needed basis. The border patrol alerts the department that it is looking for a specific vehicle. Sharing with the border country also is limited as it has diplomatic processes that must be followed to release data to the department. When this infrequent sharing is necessary, the department works informally with police in the border country.

Other Issues/Findings

- This department, as in other case studies, had one system administrator knowledgeable about information technology.
- The department said its system costs are decreasing and it can now get more with less money.
- Because local motorists know about the LPR systems, officers who pull vehicles over based on LPR hits receive less resistance from motorsts, the department said.
- The department has no policy on what action to take on system hits and leaves this to officer discretion.

APPENDIX B

Interview Protocol

Protocol for LPR Analytic Use

Thank you for agreeing to take part in this interview. I'm xxxx from the RAND Corporation in Pittsburgh, PA and I'll be facilitating the interview today.

We are conducting these interviews so that we will have a better understanding of how your agency uses License Plate Reader technology for analytical activities, such as mining previously obtained LPR data. These activities could include mining data to investigate narcotics operations, gangs, prostitution, serial crimes, pattern analysis to find stolen vehicle, as well as interagency operations like counterterrorism or border patrol activities. This is part of a study being conducted by the RAND Corporation. This work is sponsored by the National Institute of Justice.

The study we are conducting is going to evaluate the usefulness of LPR technology. This includes understanding all of the costs related to LPR, any barriers to its use or other limitations that are commonly encountered, as well as its potential benefits. While the overall study is evaluating a number of uses for LPR technology, including reactive uses, such as running license plates of cars passing by to check for outstanding warrants and stolen vehicles, we are not interested in the reactive uses for this interview. For this interview, we want to focus on more analytical uses, such as those that are initiated by mining data to investigate, for example narcotics operations, gangs, prostitution, serial crimes, pattern analysis to find stolen vehicle, as well as interagency operations like counterterrorism or border patrol activities. Here we would like you to discuss all aspects of LPR's usefulness for these analytic activities. The interview will help us better to direct the remainder of the study toward LPR's costs, benefits, and limitations. Please be honest and know that there are no wrong or right answers here.

Note that you are not required to answer any of my questions and can opt out of the interview at any time without any negative impacts. This includes questions about costs and any other questions that may be considered as sensitive information.

Protocol for Agencies that Indicated Using LPR

Overview

Your agency indicated on the survey that LPR is used here for analytical activities. Can you give me a list of the general activities you have used it for? [Make note of each activity they use it for]

[For any activity they do not mention that we have identified] You didn't mention XXXX, have you ever used LPR for this? [If no] Why not?

- We would like to learn a little bit about you, your role in the agency and in using LPR. Please tell us
  - How long have you been working at this agency?
  - What is your job title and your regular responsibilities?
  - What are your responsibilities related to LPR and how long have you been doing this?
- How often do you use LPR for analytical purposes? [If interviewee only answers "a few times," do a case study/memory jogger protocol first; if interviewee answers "many times," start with a general question and come back to case studies at the end.]

Case Studies

- Can you give me an account of the last time LPR was used for an analytical activity? Can you provide specific examples where LPR data was exploited in an analytic way to help solve an existing case or in predictive policing? [Let them tell the story and only probe to expand their description or to identify components in our model]
  - [Probes]
  - How did the database mining start? Was there an objective?
  - Did you only use this agency's database or were other agencies' LPR or other databases involved?
  - How was the data mined? How long did it take? How many people were involved in this activity?
  - Was the data mining successful?
  - [If successful] What was found?
  - How was that finding (e.g., clue) used? Did it direct the investigative strategy? How?
  - For this specific case, what were the main challenges, limitations and benefits of using LPR?
- [If the gave an account of successful story, ask for one that was unsuccessful, or vice versa] Can you give me an account where the LPR mining activity was/was not successful? [use same probes]

- Can you give me one more account of when LPR was used successfully, but maybe for a different activity? [use same probes]
  - [Identify similarities, such as part of the model, challenges, limitations, costs, etc. between stories]
  - For all accounts we just discussed, you said XXXX, would you say that is almost always the case when LPR is used for analytic activities?

Before we move on to more general questions and topics, is there anything else you want to say about these cases?

Collection

Now that we have discussed some LPR cases in detail, let's talk more generally about LPR.

Tell me about how LPR data is collected in your department. Where are the cameras placed? How many are there? What LPR equipment is used?

- What type of LPR equipment does your agency use and in what quantities?
- Are your cameras fixed or mobile?
- How many cameras?
- How many officers use LPR?
- Describe training for LPR equipment.
- How many LPR images are collected per month?
- Are there procedures, guidelines or standards that your agency follows in terms of collecting LPR data? Please describe them.
- Have you run into any problems with the LPR data collection process that you haven't already mentioned?
- Have you learned anything about LPR collection that has made the process go more smoothly? Have you changed any collection procedures in the past; if so, why?
- Can you think of any best practices related to LPR data collection that we haven't already covered?

Administration

Tell me about the administration and costs of LPR in your department. Like when did you start using LPR? How much does it cost to obtain the system and maintain it?

- Who is your department's LPR vendor?
- What was the capital investment?
- Are there maintenance, training, data storage, and/or upgrade costs?
- Are there any differences in cost between LPR administration at the local law-enforcement and LPR administration at the interagency level?

- Is the vendor responsive to problems and feedback?
- Have you run into any problems with the administration and costs of LPR that you haven't already mentioned?
- Have you learned anything about LPR administration that has made the process go more smoothly? Have you changed any administration procedures or vendors in the past, if so, why?
- Can you think of any best practices related to LPR administration that we haven't already covered?

Exploitation/Deployment

Tell me about how LPR data is stored and used in your department. Like how much data can be stored? Does your department have legal procedures for storage? Who mines the LPR data at your department?

- Where is the data stored?
- How much data can be stored in your system?
- For what length of time is the data stored?
- Are there procedures, guidelines or standards that your agency follows in terms of storing LPR data? Please describe them.
- Does your department employ intelligence analysts?
- Are any analysts dedicated to LPR data exploitation?
- What is the analytic process for using LPR data?
- Are there procedures, guidelines or standards that your agency follows in terms of analyzing LPR data? Please describe them.
- Does LPR affect strategies and/or resource allocation?
- How is LPR data linked or combined with other intelligence?
- Do investigative procedures automatically include LPR utilization?
- Does your department have access to the LPR databases of other departments or agencies?
- Are there any differences in the use and deployment of LPR between that at the local law-enforcement level and that at the interagency level?
- Are there procedures, guidelines or standards that your agency follows in terms of the linking of your department's LPR data with other databases? Please describe them.
- Do all officers have access to LPR databases?
- Have you run into any problems with the analytic use of LPR in your agency that you haven't already mentioned?
- Have you learned anything about LPR analytical activities that has made the process go more smoothly? Have you changed any analytical procedures in the past? If so, why?
- Can you think of any best practices related to LPR analytical activities that we haven't already covered?

Benefits

Tell me about the benefits you see to having LPR used analytically in your department. Like is it effective? Do you think there are crimes you wouldn't have been able to solve without it?

- What type of investigations benefit the most from LPR?
- Are any statistics kept regarding the use of LPR, its success or effectiveness?
- Number of times LPR is used? By individual investigations?
- Is it possible to rank LPR as an investigative tool compared to other investigative methods and tools?

Challenges

Tell me about any other challenges that you have not mentioned in regards to the use of LPR in your department.

Privacy

- Have you run into any privacy issues related to the LPR data?
- If yes, how has the department managed these issues?
- If no, what does your department do to ensure that there will be no privacy issues?
- Is all of the data collected and stored relevant/applicable to investigations?

Information Assurance

- Have you had any information assurance that you haven't already mentioned?
- How do you ensure:
  - The reliability and availability of the data?
  - That the correct license plate numbers are being collected and processed (e.g., image/reading problems)?

Technical Issues

- Do you have any interoperability issues related to LPR?
- Do the software, hardware, and users interface well together?
- Think about the flow/transmission of the LPR data from camera to the data storage unit.
  - Are there any bandwidth problems or other information flow problems?
  - Once the data is stored, is it easily accessed by officers in the field?
- Do you run into any problems related to sharing of data between your LPR databases and other local, state and federal agency databases?

Administration

- Do you have the manpower/intellectual capability/resources to properly use LPR as an analytic investigative tool?

- Are there any other challenges/issues with procedures for LPR use that you haven't mentioned? Are procedures in place, available, easily accessible, understandable and effectively used?

Information Flow and Interoperability Hypothesis

[Discuss the information flow and interoperability hypothesis and show flow chart] What do you think of this flow chart? Is it accurate? What is missing or extraneous?

Are there any other thoughts or comments that you would like to share before we wrap up? That's the end of my questions. Thanks very much for taking part in this discussion today; it was very helpful to us.

References[1]

29-A Maine Revised Statutes, Annotated, § 2117-A

Amar, Akhil Reed, "Fourth Amendment First Principles," 107 *Harv. L. Rev.* 757, 758 (1994).

American Civil Liberties Union (ACLU), *You Are Being Tracked: How License Plate Readers Are Being Used to Record Americans' Movements*, New York, July 2013. As of 6 May 2014:
https://www.aclu.org/files/assets/071613-aclu-alprreport-opt-v05.pdf

Angwin, Julie, and Jennifer Valentino-Devries, "New Tracking Frontier: Your License Plates," *Wall Street Journal Online*, 29 September 2012. As of 6 May 2014:
http://online.wsj.com/article/SB10000872396390443995604578004723603576296.html

Automated Plate Reader Technology, Ohio State Highway Patrol Planning Services Research and Development, Columbus, OH, 2005.

"Automatic License Plate Reader Helps Jersey Police Fight Crime," *Homeland Security News Wire*, 22 March 2011. As of 6 May 2014:
http://www.homelandsecuritynewswire.com/
automatic-license-plate-reader-helps-jersey-police-fight-crime

Baker, Al, "Camera Scans of Car Plates Are Reshaping Police Inquiries," *New York Times*, 11 April 2011. As of 6 May 2014:
http://www.nytimes.com/2011/04/12/nyregion/12plates.html?_r=1

Barnard, Patrick, "Fairfield Police Defend Use of License Plate Readers," *Fairfield Patch*, 3 June 2012. As of 6 May 2014:
http://fairfield.patch.com/articles/fairfield-police-defend-use-of-license-plate-readers

Charlotte-Mecklenburg Child Development-Community Policing Memorandum of Understanding, June 2006. As of 7 May 2014:
http://www.theiacp.org/Portals/0/pdfs/responsetovictims/RESOURCES_DOCUMENTS/4-SUSTAINING_EXPANDING/SE23_CMPD_Sample_MOU.pdf

Corillo, Todd, "License Plate Reading Earns Waynesboro National Recognition," WHSV.com, Harrisonburg, VA, 29 June 2011. As of 6 May 2014:
http://www.whsv.com/home/headlines/License_Plate_Reading_Program_Earns_Waynesboro_National_Recognition_124679249.html

Cuccinelli, Kenneth T., Virginia attorney general, letter to Col. W.S. Flaherty, superintendent, Virginia Department of State Police, 13 February 2013.

---

1 Court cases cited in Chapter Five are not included in this list.

Davis, Mark, "Vermont Bill Targets License Plate Readers," *Valley News*, West Lebanon, NH, 23 January 2013. As of 6 May 2014:
http://www.vnews.com/home/3959026-95/readers-information-plate-police

Déliberation No. 96-069 relative à la demande d'avis portant création à titre expérimental d'un traitement aautomatisé d'informations nominatives ayaant pour finalité principale la lecture automatique des plaques d'immatriculation des véhicules en movement par la société des autoroutes Paris-Rhin-Rhône (SAPR), 10 September 1996.

Drivers Privacy Protection Act, 18 U.S.C. sec. 2721 et seq.

Eisler, Ben, "ACLU Concerned Automatic License Plate Readers May Invade Privacy," WJLA.com, Washington, DC, 30 July 2012. As of 6 May 2014:
http://www.wjla.com/articles/2012/07/aclu-concerned-red-light-cameras-may-invade-privacy-78301.html

Federal Emergency Management Agency, FY 2012 Homeland Security Grant Program, web page. As of 26 April 2013:
http://www.fema.gov/fy-2012-homeland-security-grant-program

Ferraresi, Michael, "License-Plate Scanning Catching Crooks, Raising Privacy Worries," *AZCentral.com*, 23 November 2008. As of 6 May 2014:
http://www.azcentral.com/news/articles/2008/11/23/20081123autotheft1123.html

Harkins, Gina, "License Plate Recognition Databases—Good for Cops or Invasion of Privacy?" *Medill National Security Zone*, 11 May 2011. As of 6 May 2014:
http://nationalsecurityzone.org/site/license-plate-recognition-databases%E2%80%94good-for-cops-or-invasion-of-privacy/

"High-Tech License Plate Readers Aid Police but Raise Ethical Issues," *The Tennessean*, 6 May 2012. As of 6 May 2014:
http://www.wbir.com/news/local/story.aspx?storyid=219020

IACP—*See* International Association of Chiefs of Police.

International Association of Chiefs of Police, *Privacy Impact Assessment Report for the Utilization of License Plate Readers,* Alexandria, VA, 2009.

Jouvenal, Justin, "License Plate Reader Helps Nab Alleged Drug Dealer," *Washington Post*, 13 July 2011. As of 6 May 2014:
http://www.washingtonpost.com/blogs/post_now/post/license-plate-reader-helps-nab-alleged-drug-dealer/2011/07/13/gIQAOiEDCI_blog.html

Kittle, Robert, "Bill Would Ban Use of Automatic License Plate Readers in SC," WSPA.com, 2 January 2013. As of 7 May 2014:
http://www.wspa.com/story/21514683/bill-would-ban-use-of-automatic-license-plate-readers-in-sc

Klein, Allison, "Cuccinelli Limits License-Plate Cameras," *Washington Post*, 8 March 2013. As of 6 May 2014:
http://www.washingtonpost.com/local/virginia-limits-use-of-police-license-plate-cameras/2013/03/07/f1344c00-876d-11e2-98a3-b3db6b9ac586_story.html

Law Enforcement Management and Administrative Statistics, Bureau of Justice Statistics, Office of Justice Programs, United States Department of Justice, Washington, DC, 2007.

"License Plate Reader Technology Enhances the Identification, Recovery of Stolen Vehicles," CJIS Link, Federal Bureau of Investigation, United States Department of Justice, Washington, DC, 2011. As of 6 May 2014:
http://www.fbi.gov/about-us/cjis/cjis-link/september-2011/license-plate-reader-technology-enhances-the-identification-recovery-of-stolen-vehicles/

Lum, Cynthia, Linda Merola, Julie Willis, and Breanne Cave, *License Plate Recognition Technology (LPR): Impact Evaluation and Community Assessment*, Center for Evidence-Based Crime Policy, George Mason University, Fairfax, VA, 2010. As of 6 May 2014:
http://cebcp.org/wp-content/evidence-based-policing/LPR_FINAL.pdf

McKay, Jim, "License Plate Recognition Systems Extend the Reach of Patrol Officers," *Digital Communities.com*, 8 April 2008. As of 6 May 2014:
http://www.digitalcommunities.com/articles/License-Plate-Recognition-Systems-Extend-the.html?page=1

N.H. Rev. Stat. Ann. § 236:130 (2013).

———, § 261:75-b (2013).

"No License to Steal," *TECH-Beat*, National Law Enforcement and Corrections Technology Center, Gaithersburg, MD, Spring 2006. As of 6 May 2014:
https://www.justnet.org/pdf/NoLicenseToSteal.pdf

O'Brien, Luke, "License Plate Tracking for All," *Wired*, 25 July 2006. As of 6 May 2014:
http://archive.wired.com/science/discoveries/news/2006/07/71436

Office of Richard A. Brown, District Attorney, Queens County, New York, "Queens Man Pleads Guilty to Manslaughter in Death of Roommate's Friend," Press Release, 4 November 2012. As of 6 May 2014:
http://www.queensda.org/newpressreleases/2011/november/zeledon,%20luis_11_03_2011_ple.pdf

Ozer, Murat M., *Assessing the Effectiveness of the Cincinnati Police Department's Automatic License Plate Reader System within the Framework of Intelligence-Led Policing and Crime Prevention Theory*, dissertation submitted to the School of Criminal Justice, College of Education, Criminal Justice, and Human Services, University of Cincinnati, 7 July 2011.

Police Executive Research Forum, *How Are Innovations in Technology Transforming Policing?* Critical Issues in Policing Series, Washington, DC, 2012. As of 6 May 2014:
http://www.policeforum.org/assets/docs/Critical_Issues_Series/how%20are%20innovations%20in%20technology%20transforming%20policing%202012.pdf

Roberts, David J., and Meghann Casanova, *Automated License Plate Recognition (ALPR) Use by Law Enforcement: Policy and Operational Guide*, International Association of Chiefs of Police, Washington, DC, 2012. As of 6 May 2014:
https://www.ncjrs.gov/pdffiles1/nij/grants/239604.pdf

Rock, Margaret, "The Era of License Plate Tracking," Mobiledia.com, 8 November 2012.

Roper, Eric, "Cops Move to Protect License Plate Data," *Minneapolis Star Tribune*, 4 November 2012. As of 6 May 2014:
http://www.startribune.com/local/minneapolis/177120451.html?refer=y

Roper, Eric, "Man Uses License Plate Data to Repossess Car in Mpls," *Minneapolis Star Tribune*, 30 August 2012. As of 24 July 2013:
http://www.startribune.com/local/blogs/168014676.html

Roper, Eric, "St. Paul Meets Minneapolis on Vehicle Tracking Data Retention," *Minneapolis Star Tribune*, 14 November 2012. As of 6 May 2014:
http://www.startribune.com/local/blogs/179319841.html

Sengupta, Somini, "Rise of Drones in U.S. Drives Efforts to Limit Police Use," *New York Times*, 15 February 2013. As of 6 May 2014:
http://www.nytimes.com/2013/02/16/technology/rise-of-drones-in-us-spurs-efforts-to-limit-uses.html?pagewanted=all&_r=0

Taylor, Bruce, Christopher Koper, and Daniel Woods, "Combating Vehicle Theft in Arizona: A Randomized Experiment with License Plate Recognition Technology," *Criminal Justice Review*, Vol. 1, Issue 27, Georgia State University, 2011.

Urie, Heath, "Boulder License-Plate Scanner Busts 101 Parking Scofflaws in 1st Month," *Boulder County News*, 30 June 2011. As of 6 May 2014:
http://www.dailycamera.com/boulder-county-news/ci_18384625

Wilkenson, Karen, "Technology Helps Probation Department Find More Stolen Vehicles," *Citrus Heights Patch*, 8 November 2012. As of 6 May 2014:
http://citrusheights.patch.com/groups/police-and-fire/p/technology-helps-probation-department-find-more-stole86276d6ac6

Zapotosky, Matt, "Cruiser-Top Cameras Make Police Work a Snap," *Washington Post*, 2 August 2008. As of 6 May 2014:
http://www.washingtonpost.com/wp-dyn/content/article/2008/08/01/AR2008080103570_pf.html